69th Annual Meeting of the Labor and Employment Relations Association and LERA@ASSA Meeting 2017

Labor and Employment Relations Association Series

Anaheim, California, USA
1-4 June 2017

Chicago, Illinois, USA
6-8 January 2017

Editor:

Ariel Avgar

ISBN: 978-1-5108-6384-2

Printed from e-media with permission by:

Curran Associates, Inc.
57 Morehouse Lane
Red Hook, NY 12571

Some format issues inherent in the e-media version may also appear in this print version.

Printed by Curran Associates, Inc. (2018)

Labor and Employment Relations Association (LERA)
University of Illinois at Urbana-Champaign
121 Labor and Employment Relations Bldg.,
504 E. Armory Ave.
Champaign IL 61820
USA

Phone: (217) 333-0072
Fax: (217) 265-5130

LERAoffice@illinois.edu

Additional copies of this publication are available from:

Curran Associates, Inc.
57 Morehouse Lane
Red Hook, NY 12571 USA
Phone: 845-758-0400
Fax: 845-758-2633
Email: curran@proceedings.com
Web: www.proceedings.com

LABOR AND EMPLOYMENT RELATIONS ASSOCIATION SERIES

Proceedings of the LERA 2017 Meetings

LERA@ASSA Meeting
January 6-8, 2017
Chicago, Illinois

LERA 69th Annual Meeting
June 1-4, 2017
Anaheim, California

OFFICERS OF LERA FOR 2016–2017

PAST PRESIDENTS OF LERA

LABOR AND EMPLOYMENT RELATIONS ASSOCIATION SERIES

Proceedings of the LERA 2017 Meetings

LERA@ASSA Meeting
January 6–8, 2017, Chicago, IL
(in conjunction with ASSA/AEA)

and

LERA 69th Annual Meeting,
June 1–4, 2017, Anaheim, CA

Ariel Avgar, Editor-in-Chief

LABOR AND EMPLOYMENT RELATIONS ASSOCIATION SERIES
(Formerly the Industrial Relations Research Association)

Proceedings of the Annual Meeting
Annual Research Volume
Perspectives on Work Magazine

Inquiries and other communications regarding membership, meetings, publications, and general affairs of the Association, as well as notice of address changes, should be addressed to the LERA office.

LABOR AND EMPLOYMENT RELATIONS ASSOCIATION
University of Illinois at Urbana-Champaign School of Labor and Employment Relations
121 Labor and Employment Relations Building, MC-504
504 East Armory Avenue, Champaign, IL 61820
Telephone: 217/333-0072 Fax: 217/265-5130 E-mail: LERAoffice@illinois.edu
Website: http://lera.memberclicks.net EPRN: http://www.employmentpolicy.org/

CONTENTS

LERA 69TH ANNUAL MEETING, JUNE 1–4, 2017, ANAHEIM, CA

VII.
2017 LERA LIFETIME ACHIEVEMENT AWARD

LERA@ASSA MEETING, JANUARY 6-8, 2017, CHICAGO, IL

VIII.
DYNAMICS OF LOW-WAGE LABOR MARKETS: IMPLICATION FOR MINIMUM WAGE POLICY

Larry Mishel, Presiding

IX.
TOWARD A SECURE RETIREMENT: CHALLENGES AND SOLUTIONS

Charlie Jeszeck, Presiding

X.
RETHINKING PENSION FUND ACTIVISM FOR EMPLOYMENT EQUITY AND CAPITAL STEWARDSHIP

Teresa Ghilarducci, Presiding

XI.
THE ECONOMICS OF PREVAILING WAGE LAWS

Kevin Duncan, Presiding

XII.
LERA ANNUAL REPORTS

I. 2017 Presidential Address

Jobs, Opportunity, and Equality in the New World of Work: A Path Forward

JANICE BELLACE
The Wharton School at the University of Pennsylvania

For many of us, the world has changed since we were together in Minneapolis a year ago. I suppose those watching TV on election night had a sinking feeling throughout the evening. I don't know as I was in Jakarta twelve hours ahead of New York. On the morning of Tuesday, November 8, I got up and turned on the TV to CNN and immediately recognized the place—Independence Hall, in Philadelphia, my home city. It was Monday night in the United States, and Hillary Clinton was at her last campaign rally. Bon Jovi was singing, Hillary and Bill were there, and Michele and Barack Obama, and then Bruce Springsteen. There was a feeling of elation swelling up from the thousands of people on Independence Mall. With the beautiful scene of Independence Hall lit up, I switched off the TV to get to the airport. When I was in the airport in Singapore it was now morning in America and people were going to the polls. Then I got on the plane for the 13-hour flight to London. The flight got in at 6 am, 1 am in New York. I got my bags and went out to meet my driver, and asked Abdul "is there any news yet about the American election?" And he said, "The man won." I said, "What??" And he said, "Trump. He won. How could America have done this?"

Was I shocked? Not really. Surprised, yes. But not shocked. Not shocked because undertones of what was going to come were sounded months earlier. In July, when I was on vacation with very good friends from college, there had been an article in the Financial Times, "Trump's message shines in faded steel city,"[1] days after Trump had visited the town. It was about Monessen, Pennsylvania, a place most of you probably have never heard of. A steel town where the mill had closed, which caused the tax base to drop dramatically, which led to abandoned homes and blight, where, as the FT noted, there was "nowhere near enough jobs to support the population." But we knew Monessen, a steel mill town between the West Virginia border and Pittsburgh. We knew because one of us, Larry, came from Monessen. His father had worked at Wheeling Pitt, and at the point it went bankrupt around 2000, Larry's father had a major heart incident and needed hospitalization but the workers had been stripped of their health insurance. As the FT article observed, this town had always voted Democratic but people kept telling their reporter "no one is listening to us."

Just before the election, the Financial Times reporter went back to Monessen and talked to the 78-year old mayor, a lifelong Democrat talked about the people who felt they had been taken for granted, and how the Clinton and Obama had never even responded to letter he sent.2 Noting the town's desperate economic plight, he said: "If ISIS was to come to Monessen, they'd keep on going. They'd say someone already bombed the goddamn place."

What was the 2016 election about? Rage. Anger. Despair. With Trump the vessel expressing this for those not heard, those taken for granted.

Rage Against Job Losses

There was rage against job losses, often expressed as opposition to free trade agreements. But it was liberalized international trade, not FTAs *per se*, that was the cause. Compared to the 1990s, by 2016 there was widespread recognition that trade creates winners and losers—with manufacturing workers in the higher

Author's address: Wharton School, 672 Jon M. Huntsman Hall, 3730 Walnut Street, Philadelphia, PA 19104

> wage country often losing. There is more than one reason for the job losses in manufacturing over the past 25 years, but it is undeniable that international trade is one major reason. As David Autor and his colleagues have found in a study of communities competing with China on specific products, "… rising imports cause higher unemployment, lower labor force participation, and reduced wages in local labor markets that house import competing manufacturing industries" which in turn led to "a steep drop in the average earnings of households."[3]

What was expressed was rage against job losses. Lost because workers abroad could do the work so much more cheaply. Lost because the job has been eliminated by some technological advance. And, most importantly, because the workers had no new good jobs to go to.

Today there is widespread realization that raising skill levels of workers will take substantial time, and that most workers over 50 who lose jobs will never be able to regain their place in the labor force. Moreover, over the past 50 years, an era of deindustrialization, the United States has not demonstrated an ability to re-skill workers who lose their job due to offshoring of production or technological innovation. The result? As Angus Deaton and Anne Case found rising mortality;[4] in other words, "Death by Despair" (in the more memorable words of the magazine article that reported on these findings[5]). The increase in mortality for middle-aged (45-54) white males is shocking for another reason: the United States is the only advanced country that has experienced rising mortality for majority group, prime age males.[6]

But the outlook is even worse. Even if higher wage countries raise the skill level of workers, higher tech industries employ fewer persons. There is anxiety today that the supply of educated persons may be higher than current demand in high wage service industries. Rather we have what I call "the barista with a bachelor's degree" syndrome.

We have seen anger, rage, despair, and sometimes resignation. Even when emotions are not this high, we know that there is a very high level of anxiety in the country about our future.

Who Can Make America Great Again?

What has gone wrong? Who can fix it? Who can make America great again? Such questions should make us ask another—*when* was America great?

Do we think of the 1950s? For those of you who have seen the recent film, "Hidden Figures" (about three brilliant African-American women who worked at NASA when John Glenn became the first man to go into orbit), maybe you will answer 1962—when we were not polarized but united as a nation.

Why Was America Great?

The United States has a very large domestic market, and it is blessed with many natural resources. It might also be that in the late 1950s most of our competition was rebuilding after a devastating war.

But ask, was our labor relations system sound and sustainable? We think back to an era with a high rate of unionization—35%; an era when management and labor more or less cooperated; and with real gains in wages and benefit levels.

But What Was the Basis of This Apparent Stability in Labor Relations?

- Lack of a perceived need on the part of managers to compete on labor costs
- Lack of much competition

When did it change? And why?

On various metrics, the change begins in the 1970s. Look at percent unionized, real wages, amount invested in education;[7] amount (as a percentage of income) saved in pension funds. The line moving upwards flattens in the 1970s or started to go downwards. I noticed this when writing an overview article.[8] I took the view that the oil crises of the 1970s, and the resulting stagflation, caused this. But did it? Now I think I got it wrong. The oil crises and stagflation were not the primary cause.

Other countries, such as the Scandinavian countries, Germany, faced the same oil crises. But they did not experience what we experienced. Why? Because they responded differently.

Changing View of the Purpose of the Corporation

Consider the prevailing view prior to the 1970s. Throughout the 1950s and 1960s, federal expenditures were growing at a quick pace in the areas of national defense, social welfare, and infrastructure. Both major parties, Democratic and Republican, supported increased spending in different ways. This reflected the acceptance of Keynesian economics. But then, a change occurred. The early signs came in 1962, when Milton Friedman published "Capitalism and Freedom."

In 1970, Milton Friedman authored a very influential article that was published in the *New York Times* Sunday magazine, "The Social Responsibility of Business is to Increase Profits." He bluntly stated that the business of corporate managers is to make money for the shareholders, and not for other purposes unless required by law. The message was clear: other stakeholders don't count—only shareholders.

Related to this, in a more practical fashion, was the 1976 paper by Michael Jensen and William Meckling in the *Journal of Financial Economic* that established shareholder value; that is, they argued that the shareholders were the principal and management was the agent, and that the ultimate measure of a company's success is the extent to which it enriches shareholders.[9]

This was theory. To apply theory to practice, one had to consider how managers behave. This led to articles examining how to align the interests of managers with those of shareholders; in other words how to incentivize executives to maximize shareholder value. The obvious way to incentivize executives is to make their compensation dependent on maximizing shareholder value, euphemistically expressed as "aligning executive compensation with shareholder interests."

Recall that in 1970 stock-based compensation represented less than 1% of CEO remuneration. Then look back to the 1980s, a decade of stock repurchases, corporate restructuring, and green mail—all designed to maximize the value of shares.

How did the top management of companies respond? They felt compelled to cut costs. How?

So many costs were out of the control of management. The easiest and fastest way to cut costs was to cut labor costs.

How was this achieved? Even before the 1980s, it was evident that one way to cut labor costs was to move production to nonunion, lower wage areas. In the 1950s, the garment industry moved shops from the North to the South (the runaway shop cases). By the 1970s, the garment industry, the most labor-intensive industry, started leaving the country. Even before free trade agreements, the wage difference between Asia and the U.S. was so great that despite tariffs it was still cheaper to manufacture abroad. (I know—in 1975 the men's clothing factory where my father had worked for 30 years closed as it was so much cheaper, with the same low level of technology, to make men's trousers in Asia.)

Then this strategy extended to other industries. Recall the "Southern Strategy" of the major auto companies; namely, open plants in the nonunion South and stop investing in the heavily unionized North and Midwest.

Other strategies for lowering labor costs were utilized, such as

- break unions
- "hard bargaining" (a union victory but no first contract)
- decertification
- undermine pattern bargaining (e.g., airlines)
- concession bargaining

If we look back at the titles of papers and sessions at IRRA meetings in the 1980s, all of these pop up.

The Weakness of American Labor Law

In the late 1970s/1980s, the weakness of American labor law became depressingly apparent. One only has to recall the most basic labor economics reality to understand why.

John R. Commons once observed that a union must organize the length and breadth of the market. That is, a union that has *organized* ***all*** *the workers* producing *the* ***same*** *product for the* ***same*** *market* can seek to raise wages. Since employers compete with each other, for one employer to remain competitive with others in the same industry (producing the same product), that employer has to keep its costs in line with its competitors. The basic economic fact is that a union must take labor out of competition in order to impose higher labor costs.

In line with John R. Commons' observation, it is clear that a union should seek industry-wide bargaining. In Europe, in many countries, industry-wide bargaining was the norm. In tough times, those unions could rely on that to retain their ability to bargain, as companies competing in the same industry with each other were all bargaining together. This is why German and Swedish and Danish unions were able to withstand the tough economic times of the 1970s.

In the United States, it was not that the oil crises of the 1970s radically increased the cost of production that *caused* American unions to decline (if so the result would have been the same in Germany). It was that American management was now under pressure to increase shareholder value *and at the same time,* costs that management could not control were going up sharply. Thus, the management of companies was under pressure to cut costs where they could—labor. And there was no bulwark against this.

American labor law never had supported *industry-wide* bargaining, but worse, American labor law contemplated bargaining at a lower level (the "employer"—the company). Even worse, judicial decisions from the 1950s–1970s effectively pushed the level even lower by permitting employers to insist on an election even in the absence of any good faith doubt. To win an election, a union must have 50% plus one in a unit. We all know how difficult it is to organize a multi-site company all at the same time and to go to an election on the same date. Thus, elections were taking place in parts of a company, not the entire company (let alone an industry). The result—companies in negotiations could simply play one plant off another (if you want to keep the work here, agree to these concessions).

The Need to Rely on Strikes

The other fundamental weakness of American labor law relates to the fact that it is totally dependent on strikes. Unlike other countries, there are no works councils, no rights to information and consultation. In other words, workers have the right to engage in collective bargaining, but their ability to bargaining collectively to beneficial effect is 100% reliant on strike power. Yet, American labor law has *never* protected strikers.[10]

In the 1970s, when employers began to engage in hard bargaining, and workers had to strike to hold on to what they had, they faced grim possibilities:

- Strike—and perhaps be permanently replaced
- Strike—and see the strike fail, and come back having gained nothing and lost money
- Strike—and get a new contract but see that over time no new investment is made in the plant and work is going to new nonunion plants

It is no wonder that workers who were interested in unions (as surveys repeatedly revealed) declined to vote for unionization (after having been repeatedly reminded by the employer that the only way a union forces the company to give more than it intended is to strike, and that strikes could have dire consequences for the workers).

Can labor law be reformed? Twice, under Democratic presidents (Carter and Clinton), this has been tried. It has failed. There simply was not enough support in Congress.

A Limited Statute Now Outdated

But even those labor-law reform efforts, if successful, would not have produced the desired result because the NLRA is a fundamentally flawed statute, designed to fix the problem of the moment, and not helped by the fact that the courts have interpreted it in a very narrow fashion.

Thirty years ago, I presented a paper at the 1987 IRRA annual meeting entitled: "Mandatory Consultation: The Untraveled Road in American Labor Law." In the paper, I argued that the NLRA did not mandate that adversarial bargaining on a limited number of subjects was the only protected right. I pointed out that it was not until 1958 that the Supreme Court, in Borg-Warner,11 created three categories of bargaining (mandatory, permissive and illegal subjects). I argued that something was left out: mandatory subjects of consultation.

I argued that even under the *Borg-Warner* construct, there were Items on which employees could not strike but that they were entitled by virtue of their Section 7 rights to receive information and to engage in consultation (such as the company's plans to invest in the plant).

I recall the reaction of a colleague after hearing this presentation. He said:

> Janice, why make this proposal? Who would support such a change? Employers like the NLRA—they know they can beat unions. And unions don't want this—they're afraid it will give workers some other option for voice and that will be enough to satisfy them.

I recall that at the meeting itself, someone in the room voiced this last sentiment, and I said, "Union membership is down to 18%. How much lower do you want it to fall?"

In 2016, there were 14.6 million members in the U.S.—down from 17.7 million in 1983.

The percentage of workers belonging to a union was 10.7%, compared to 20.1% in 1983.

And these figures somewhat disguises the extent of the decline (because of changes in the relative proportion of private to public sector unionism, the size and composition of the labor force, etc.). Today, in the public sector, 34.4% are unionized—more than five times higher than in the private sector. Union membership in the private sector has fallen to 6.4 %—levels not seen since 1932. A figure that even more starkly points to the decline is the fact that there are now more undocumented workers in the U.S. than unionized private sector workers. It is difficult to conceive of a revival under our current labor laws. With the decline of manufacturing, with the rise of automated service sector work, how many workers have strike power and are willing to take the risk of going on strike?

I have painted a gloomy picture. So where do we go from here? Are there any hopeful signs? I think there are.

It may be in the growing and now widespread realization that over the past 40 years, gains have not been shared. The popular slogan is "the 1 %." The point is that many people in America today realize that shareholders have walked off with the lion's share of the gains.

Another hopeful sign is the emerging reappraisal of shareholder value theory because an awareness of how this theory has affected society might lead to a demand that other management consider other stakeholders, such as those who invest their human capital in the firm.

A signed article in 2011 in *Forbes* had the catchy title "The Dumbest Idea in the World: Maximizing Shareholder Value."[12] This view was echoed by Lynn Stout, of Cornell Law, in her 2012 book, *The Shareholder Value Myth: How Putting Shareholders First Harms Investors, Corporations and the Public.* Most encouraging is Duff McDonald's book, which was published in April, entitled *The Golden Passport: Harvard Business School—The Limits of Capitalism and the Moral Failure of the MBA Elite.*[13] I say "most encouraging" because McDonald directly confronts the public policy issue; namely, is it moral for shareholders and top executives to capture the lion's share of the gains many have worked to create.

In a surprisingly favorable long book review published in the *Wall Street Journal*[4] that supported McDonald's critique of shareholder value, McDonald is congratulated for producing a "guidebook" and a "deliciously iconoclastic history of the Harvard Business School" for those who "sift through the wreckage of a civilization that bestowed its highest rewards on individuals trained to ignore its deepest problems."

There is another hopeful sign. People are recognizing that for a society to be healthy, gains must be shared. But there is absolutely no consensus on what this means, nor how it can be accomplished. And at present, we cannot look to Washington.

A Path Forward

If there are hopeful signs, I must also point out some dark clouds.

Our economy is at the point of a paradigm shift. Industrialization is over. Models from the Industrial Age are likely obsolete. We are firmly at the beginning of the Digital Age. We saw the early signs and now we see a huge sign.

Consider the changes in the nature of work since the 1970s.

- **1974:** Intel 8080 microprocessor factories CAD/CAM—jobs disappear
- **1974**: Barcode scanner first used—"tracking" jobs disappear (retail stores, deliveries)
- **1993:** World Wide Web—routine office jobs disappear (clerical, counting)
- **2020:** Artificial intelligence—*from* many people doing rote tasks (manual or mental) *to* fewer people using their minds

This huge movement into a radically new economy has re-surfaced patterns from a century ago—that capital used labor, not paying for more than it needs.

- **Crowdsourcing**
  - 19th century: workers swarming down to the docks when a ship came in
  - 21st century: workers logging on to the platform, bidding to work
- **The gig economy**
  - 19th century: musical performers, day laborers
  - 21st century: TaskRabbit, Amazon Mechanical Turk
- **Matching demand for labor with supply of labor**
  - 19th century: labor supply broker owner's agent
  - 21st century: technology platform (e.g., Uber, Lyft)

At this particular point in time, in this political climate, it is not realistic to call for labor-law reform, to propose policies that would call for substantial government funding (more for education, be it pre-school or college).

It is possible to consider approaches that worked over a century ago—mobilizing workers with a focus on issues and pressing for change. What is one of the most successful, most broad based associations in the United States that is not a political party? The AARP—the American Association of Retired Persons. Maybe we need the *AAWP—the American Association of Working Persons.* The AAWP

- would propose and rally support for changes
- would not be based on traditional "employment"
- would be linked to individuals and 'units' of work

The AAWP would focus on issues that concern working persons such as

- Support for higher education/training
- Home ownership plans

- Long-term savings for retirement
- Insurance for sickness, disability

Are these labor organizations? Not the way we have known them.

Nor are they labor organizations within the meaning of the NLRA because they are not linked to one employer or even many employers in one industry. Moreover, they are not seeking to bargain collectively with an employer.

A year ago, before the Democratic and Republican national conventions chose their candidates, I chose the theme of this annual meeting: Jobs, Opportunity and Equality in the New World of Work. As the popular response to the film "Hidden Figures" demonstrates, a film where people worked hard and merited the job, people in 1962 and people today want jobs, they want opportunity, and they want equality—an equal chance at the jobs available.

What I propose is not a union. Not the way we have known it in the Industrial Age.

But today's unions should support the AAWP because by mobilizing working persons to support policies that aim to benefit those who work in America and their families, we will be on the path to moving forward to achieve *Jobs, Opportunity and Equality in the New World of Work.*

Endnotes

[1] Subtitle: "Formerly Democratic Monessen is won over by insurgent Republican," by Demestri Sevastopulo. July 1, 2016. http://on.ft.com/2k6xfTg

[2] "How Trump gave a voice to unheard America" by Demetri Sevastopulo. *Financial Times*, October 27, 2016. http://on.ft.com/2k6k5W6

[3] David H. Autor, David Dorn, and Gordon H. Hanson, "The China Syndrome: Local Labor Market Effects of Import Competition in the United States." *American Economic Review* 103:6 (2013) 2121-2168. http://bit.ly/2k6c97i

[4] Anne Case and Angus Deaton, "Rising morbidity and mortality in midlife among white non-Hispanic Americans in the 21 century." Proceedings of the National Academy of Sciences. Sept. 17, 2015. http://bit.ly/2k6yYrG. See also, Anne Case and Sir Angus Deaton, "Mortality and morbidity in the 21 century," Brookings Papers on Economic Activity, Spring 2017. http://brook.gs/2k6zhTm

[5] Olga Khazan, "Middle-Aged White Americans are Dying of Despair," *The Atlantic*, Nov. 4, 2015. http://theatln.tc/2k9fUJx

[6] Ibid. A chart showing this in reproduced in the *Atlantic* article, with the chart taken from Case and Deaton's article in the PNAS (see endnote 4).

[7] Claudia Goldin and Lawrence F. Katz, *The Race between Education and Technology*. Cambridge: Belknap Press of Harvard University Press (2008).

[8] Janice R. Bellace, "American Unions and the Economy: The Unheard Voice of a Shrinking Sector," *Singapore Economic Review*, 59:4 (Sept 2014) 1120. http://bit.ly/2k6cMha

[9] Michael C. Jensen and William H. Meckling, "Theory of the Firm: Managerial Behavior, Agency Costs and Ownership Structure," *Journal of Financial Economics*, 3:4 (1976) 305-360. http://bit.ly/2k8fcvN

[10] *NLRB v. Mackay Radio & Telegraph Co.*, 304 U.S. 333 (1938). One year after the National Labor Relations Act was held constitutional, the Supreme Court in a 7-0 decision held that workers who strike remain employees but observed that they may be permanently replaced without any requirement that the employer demonstrate any need to do so.

[11] *NLRB v. Wooster Div. of Borg-Warner Corp.*, 356 U.S. 342 (1958).

[12] Steve Denning. *Forbes*. Nov. 28, 2011. http://bit.ly/2k7gV4N

II. AILR/LERA BEST PAPERS

Workforce Training for Older Workers: Toward a Better Understanding of Older Worker Needs After the Great Recession

TING ZHANG
University of Baltimore

Facing the aging workforce but older workers' vulnerability in the labor market, this paper empirically explores factors and policy implications to enhance older workers' employment propensity, measured by the entered employment rates (EER), after exiting the national workforce program. After reviewing older workers' attributes and the unique methods to train them, the paper examines effects of program attributes and local labor market cyclical changes on older workers' EER, controlling for individual workers' demographic and socioeconomic characteristics. The paper relies on three types of empirical models including simple logistic regression, mixed-effects regression, and multilevel mixed-effects logistic regression models for robust estimates. The models are conducted separately among older dislocated workers and among older adults. Longitudinal 2013-2015 Workforce Investment Act Standardized Record Data (WIASRD) and Bureau of Labor Statistics unemployment data are used. Some WIOA training and related service combinations are identified to contribute to older adults and older dislocated workers' EER and to inform strategic decision-making about future allocations of funds and policy efforts to serve older workers. For example, on-the-job training greatly enhances both older dislocated workers' and older adults' EER. While skill upgrading training and supportive services only enhance older dislocated workers' EER, participating in mechanical, transportation, and military skill trainings only enhances older adults' EER. While older adults' EER are affected by local economic downturns, it is not the case for older dislocated workers. The study concludes with policy implications and future study directions.

Introduction

Older workers are vulnerable in the labor market because they are often "pushed" out of the traditional employees' market, particularly in the Great Recession (Lavarreda, Snyder, and Brown 2013). One "push" factor is limited training and promotion opportunities, in addition to factors like age discriminatory practices in recruitment, a lack of attractive employment options, deteriorating health issues, and care responsibilities (Porcellato et al. 2010). Older workers are generally perceived to be subject to skill obsolescence (Crown and Longino 2000). For older workers who have not looked for a job for a long time, lacking job-hunting techniques is another structural disincentive for boomers (Hooyman and Kiyak 2005). However, older workers' valuable experiences, work ethic, mentoring, and language skills are also recognized (Collison 2003). Continue working can provide older workers with an additional income in retirement while they remain economically active with flexibility and independence (Weber and Shaper 2004; Kautonen et al. 2008; Zhang 2008, 2014, 2015; Zhang and Carr 2014). It is therefore critical to deliver an appropriate level and mix of workforce services for older workers who need training to continue working or to return to work.

Author's address: Department of Finance and Economics, Merrick School of Business, University of Baltimore, 1420 North Charles Street, BC-383, Baltimore, MD 21201-5779

Older workers' demand for training to remain or regain labor market competitiveness is certain to increase. This has policy implications for the Workforce Innovation and Opportunity Act (WIOA) planning. This research uses the most updated data and multilevel mixed-effects modeling to examine and help inform investments in future workforce services for older worker training. Relying on the Workforce Investment Act Standardized Record Data (WIASRD) 2013-2015 quarterly microdata and Bureau of Labor Statistics (BLS) unemployment data, this paper empirically examine the impacts of unemployment rates and various workforce services on older workers' entered employment rate (EER),[1] controlling for older workers' demographic and socioeconomic conditions. Some WIOA training and related service combinations are identified to inform strategic decision-making about future allocations of WIOA funds to serve older workers. Following the finding interpretation and discussions, this paper concludes with policy implications.

Literature Review

Older workers' share in the labor force is rising quickly. According to the Bureau of Labor Statistics (BLS) (2017a) data, currently 23% of the US total employment are from those who aged 55 and above, based on calculation. In 2014, the 21.7% of the labor is for ages 55 and above, up from 11.9% in 1994, 15.6% in 2004; this number is projected to increase to 24.8% in 2024 (BLS 2017b). The actual number of older labor force participants could be even greater than official projections suggest (Rix 2006). Concerned about the aging labor force, prior studies have addressed older workers' vulnerability and assets, potential training methods for older audiences, and available resources related to the training.

Older Workers' Vulnerability

Older workers are particularly vulnerable in the economy. Compared to younger workers, older workers are increasingly more likely to be laid off, are less likely overall to find another job, take longer to find a new job, and tend to earn less income after finding a job (Heidkamp 2009).

Many employers are concerned about the higher costs and lower productivity of hiring older employees. The higher cost concern include concerns about higher health insurance and benefit costs (Collison 2003). Even the "Seniority Principle" (Thurow 1975) which increases salary with experience tends to discourage hiring seniors. For the lower productivity concern, older workers are generally perceived to be more vulnerable to skill obsolescence and tend to be at odds with technological innovations that are associated with increased training needs (Crown and Longino 2000). For those older people who have not been searching for a job for a while, lacking job-hunting techniques is another structural disincentive for older people (Hooyman and Kiyak 2005).

There is also general doubt about older workers' trainability as well as employability. Many cognitive abilities such as working memory, the capacity to pay attention, and spatial cognition decline with age, especially when a task is complex or represents an unfamiliar knowledge/skill area (Charness and Czaja 2006). In addition, older workers have a shorter work life left to recoup the costs of training.

Older Does Not Mean Less Productive

Older workers may have more difficulty than younger ones for learning new knowledge, but it does not necessarily mean they are less productive. People's intelligence is composed of crystallized intelligence and fluid intelligence. Crystallized intelligence is a form of acquired knowledge and is usually stable until very late life; fluid intelligence refers to ability to quickly solve novel problems and shows declines from the 20s or 30s (Schulz and Salthouse 1999). Although older workers may have a disadvantage in fluid intelligence, they may not have a disadvantage in crystallized intelligence. For example, older workers tend to have better language skills.

Actually, various studies show that age is a poor indicator of work performance and that variations in performance within the same cohorts far exceed the average differences between cohorts (Sterns and McDaniel 1994; Human Resources Development Canada 1999). Using data on U.S. General Motor employees, the Florida State University Psychology Department and the Pepper Institute on Aging and Public Policy also found that older workers were not less productive or valuable in the workplace, despite their longer learning processes (Charness 2004).

Even for new skills related to high technology, research also shows that older people's deficiencies in computer skills, for example, are really a function of socially driven motivation (Friedberg 2003; Resnick et al. 2004). Social factors may be more important in modulating motivation to train (Colquitt et al. 2000; Maurer et al. 2003).

Although marketing and training for new information and communication technologies has focused mainly on younger people, some studies (such as Resnick et al. 2004) demonstrated that older adults could learn to use the Internet, and that Web use could improve their quality of life. A number of other studies (e.g., Elias et al. 1987; Gist et al. 1988; Czaja et al. 1989; Charness et al. 1992; Morrell et al. 1995; Mead et al. 1997) have examined the ability of older adults to learn to use a variety of computer applications with perspective training strategies. Based on those studies, there were no discernable differences in performance for different age groups at the end of training, and performance was often better among older novices than that achieved by younger novices, though the older group requires more time.

Older Workers' Valuable Assets

On the other hand, older workers can bring many unique qualities to the workplace. The wisdom of older people is typically represented by experience, guidance, leadership, and comfort (Peterson 1999) and older workers typically bring maturity, dependability, and years of relevant experience to the workplace (Eyster et al. 2008). As identified by the 2003 SHRM/NOWCC/CED[2] Older Workers Survey, older workers have invaluable experience, established business ties, a strong work ethic, loyalty to their company, diversity of thoughts and approaches, and can provide additional support to younger workers through mentoring (Collison 2003).

Older workers possess unique skills that younger workers might not have. Older workers can also be effectively trained. Facing aging and pending workforce shortage,[3] job training that helps older workers obtain and succeed in a job could play a necessary and pivotal role in workforce development and economic growth.

Methods to Train Older Workers

Understanding older workers' vulnerability and values would help to understand training methods that are suitable for older workers. Previous literature addresses some clues on how to train older workers. Generally, what is best for young adults works for older adults; however, specific approaches could be particularly helpful for seniors.

Previous knowledge has been found to be a strong and positive predictor of learning performance both for in-class instruction and homework learning activities (Beier and Ackerman 2005). This mirrors Schmidt and Hunter's (2004) research that indicated experience as a better predictor of job performance than intellectual abilities. Therefore, training based on their previous job experience and skills could be effective.

Certain physical or other accommodations can be made to enhance older workers' productivity. With seniors' decreasing vision, high contrast settings for print or computer screens can effectively stimulate their vision; similarly important includes setting adjustments for audio equipment (see Fisk et al. 2004). Previous literature also found that there were greater gains for older adults when performing procedural (action or hands-on) activities, compared to conceptual training using automated teller machines (Mead and Fisk 1998). A similar differential advantage was shown on training to search the Internet (Mead et al. 1997). Moreover, matching the instructional technique and medium (e.g., text, voice, animation) to the type of material that is being presented might be helpful. Older adults might have greater difficulty ignoring irrelevant information than younger adults based on basic experimental work on attention spans (Carlson et al. 1995); though Schneider, et al. (2000) disagree.

Seniors are typically slower to acquire skills than younger adults, though some of the slowing in learning may be explained by older adults' preference for accuracy over speed, with the reverse holding true for younger adults. Charness and Czaja (2006) indicated the importance to allow extra time to train an older adult (1.5 to 2 times the training time expected for a young adult).

In addition, many older workers lack job-hunting techniques (Hooyman and Kiyak 2005). Therefore training on job searching, hunting, and interviewing skills could be useful as well.

Older Workers' Training Resources

Publicly funded training programs have been serving older workers for decades. The Workforce Investment Act (WIA), the federally funded employment and training program, provided skill upgrades to workers regardless of age, and allowed states and local areas to give priority to special populations, such as older workers, for training funds allocation. Some states began to focus on older workers' training needs and to tailor some services and funding to serve older workers. Programs funded through WIA served a growing number of older workers.

On July 1, 2015, Workforce Innovation and Opportunity Act (WIOA) took effect. Written into law on July 22, 2014, WIOA supersedes the WIA of 1998 and is designed to help job seekers access employment, education, training, and support services to succeed in the labor market and to match employers with the skilled workers they need to compete in the global economy (Employment and Training Administration [ETA] 2017a).

Shown in Table 1, according to the Workforce Investment Act Standardized Record Data (WIASRD) Data Book, comparing across Program Years (PY) 2013-2015,[4] among all WIA/WIOA program exiters, adult workers aged 55 and above[5] increased from 14.8% in 2013 to 15.2% in 2014 and to 15.8% in 2015; and dislocated workers[6] are aged 55 and above increased from 21.5% in 2013 to 21.3% in 2014 and to 22% in 2015 (ETA 2017b). As the population ages and with the elevated financial pressure from the Great Recession, the use of WIOA-funded training by older workers will likely continue to grow.

TABLE 1
Percentage of Older Workers (55+) Among all WIA/WIOA Program Exiters, PY 2013-2015

	Program Year		
WIA/WIOA Exiter	**2013**	**2014**	**2015 (4/1/15–3/30/2016)**
Older Adults	14.8	15.2	15.8
Older Dislocated Workers	21.5	21.3	22

Data sources: ETA (2017a).

In addition to the WIOA (or previously WIA) funding, the U.S. Department of Labor (DOL) funds Senior Community Service Employment Program (SCSEP) sites, a dedicated employment and training program offering job search assistance, training and work experience to lower-income seniors. SCSEP provides many grants to state governments and national nonprofit organizations to train older workers (ETA 2017c). Some community colleges are also leading efforts to develop job-training opportunities for older workers to meet local labor market needs.

Research Hypotheses

With a rising share of older workers enrolled in the federally funded WIOA programs, what level and mix of the WIOA employment and training programs will be needed to serve older workers? When investigating EER, it is also necessary to determine if the WIOA program effectiveness is sensitive to cyclical changes in labor market conditions considering the heterogeneity of demographics, socioeconomic background, location, time of enrollment and even occupational preferences.

While prior literature on this investigation is thin, this study addresses these above questions and test the following two hypotheses:

- Some WIOA training and related service combinations can be identified to inform strategic decision-making about future allocations of WIOA funds to serve older workers.

- WIOA program success with older workers is sensitive to cyclical changes in labor market conditions.

The following sections introduce the methodology adopted in this study and then present observations based on descriptive statistics and empirical models. Findings are discussed and policy implications are indicated where appropriate.

Methodology and Data Sources

This paper adopts simple logistic regression, mixed-effects regression, and multilevel mixed-effects logistic regression models to test the impacts of unemployment rates and various WIOA program attributes on older workers' EER. WIASRD 2013-2015 data and BLS unemployment data series are used. This research starts with descriptive statistics.

The WIASRD 2013-2015 data are used as a longitudinal dataset. The temporal unit of analysis is the WIOA program exiting date for each participant in a state. For the multilevel hierarchical modeling, the two higher levels' units of analysis include Workforce Investment Board (WIB) areas (also called Workforce Investment Areas) for the middle level and state for the highest level. Post-estimation regression diagnostics is conducted to address the model fit and specifications. Older workers in this study are defined as workers age 50 and above. This age divide is often used as a definition for older workers and represents the front edge of an older worker cohort with employment problems that are generally regarded to be particularly acute (Zhang 2014). Compared to another often-used age definition for older adults, 55, starting with the age of 50 could include many more older workers and include more candidates who can benefit from the older worker workforce program designs. The range of 50-55 years old is an important age span for workforce program participation that prepares for older workers' continued working life.

Dependent Variable

The dependent variable is **Enter Employment Rate** (EER), consistent with the common measure EER of the WIOA program performance measure. For the WIOA adult program, of those who are not employed at registration, EER is defined as the number of adults who have entered employment by the end of the first quarter after the exit quarter divided by the number of adults who exit during the quarter. For the WIOA dislocated worker program, EER is defined as the number of dislocated workers who have entered employment by the end of the first quarter after the exit quarter divided by the number of dislocated workers who exit during the quarter.

The WIASRD data are extracted to measure the EER for an **individual** *(i)* and for a **specific program exit date** *(t)*. Since we use the individual level microdata, this measure is a binary variable, the logarithm of which actually measures the entered employment propensity. This reflects the WIOA program common measure EER at an aggregate geographic area level. If the individual participant found a job during the 3 months after exiting the program, the value of *EER* is 1; otherwise, it is 0. Although using EER as a measure for WIA/WIOA program performance is not perfect,[7] EER is still a legitimate and best available direct measure when this research is conducted. This one-quarter lagged value allows a quarter's time for workers to find a job after exiting WIA/WIOA programs and services; this also avoids simultaneous causation and endogeneity issues in statistical inference. In this study, the WIASRD data used for the regression models covered exit dates ranging from April 1, 2011 through June 30, 2016.

Independent Variables

The first set of independent variables include program types, training types, and occupational categories of training. Those variables are used to test the first hypothesis and observe what program, training, and occupational category mix works the best for older workers. Those variables are binary variables with 1 for the referred categories and 0 otherwise. The **program types** ***(P)*** include the following: *Supportive Services (except needs-related payments)*, *Needs-Related payments*, the *National Emergency Grant* program, *Pell grants*. The **training types (*T*)** include *On-the-Job Training*, *Skill Upgrading and Retraining*, *Entrepreneurial Training*, *Adult Basic Education (ABE) or English as a Second Language (ESL) in Combination with Training*, *Customized Training*, and *Other*

Occupational Skills Training. The **occupational categories of training** ***(O)*** include (1) *Agricultural, forestry, fishing and related workers, construction and extractive workers,* (2) *Managerial, administrative, professional or technical,* (3) *Sales, clerical and administrative support,* (4) *Service workers,* and (5) *Mechanics, installers, repairers, precision workers, machine setters, set-up operators, operators, tenders, assemblers, hand workers, transportation and related workers, and military.* The information for these variables is extracted from the WIASRD data.

Another independent variable is state level monthly **unemployment rate** (*U*). It captures impacts of regional cyclical labor market conditions and tests the second hypothesis. This is a continuous variable. The BLS non-seasonally adjusted unemployment rate data are used for this variable. Considering the limited sample size from the Current Population Survey, the BLS Local Area Unemployment Statistics (LAUS) unemployment rate information is more reliable at the more aggregate state level instead of at the county level. This paper adopts monthly unemployment rates at the state level to observe the business cycle impact. It is impossible and unnecessary to monitor unemployment rates daily.

Control Variables

The WIASRD data also offer older information on workers' background, including demographic, socioeconomic, and other program participation information details of older workers who have participated in WIA/WIOA programs. The **demographic** and **socioeconomic** information includes gender, age, disability, education attainment, veteran status, homeless status, whether the participant had limited English skills, and whether the participant received services or assistance from the Temporary Assistance to Needy Families (TANF) program, the Supplemental Nutrition Assistance Program (SNAP) program, and/or the Supplemental Security Income (SSI)/Social Security Disability Insurance (SSDI) under Title XVI of the Social Security Act. English skills are often needed at work. The homeless status and participation into the TANF, SNAP, SSI, and SSDI programs are often associated with poverty for different reasons.

The **WIA/WIOA program participation information** includes whether the participant received financial services/assistance under the Wagner-Peyser Act, the Local Veterans Employment Representative (LVER) Program and Disabled Veterans Outreach Program (DVOP), the Older Americans Act of 1998, or the Carl D. Perkins Vocational and Applied Technology Education Act, or received training or received services through distance learning, supportive services, SNAP Employment Training program (SNAP E&T), a Pell Grant, Trade Adjustment Assistance (TAA) programs, the American Recovery and Reinvestment Act (ARRA) of 2009, needs-related payments, apprentice training, other basic skill training, or a National Emergency Grant (NEG).

With program participation date and exit date, this paper computes **program participation length (*L*),** to capture how long it took a participant to finish the program. The program participation length is determined by the program design as well as individual motivations, efforts, and access. This variable is believed to affect an individual's EER.

Please note that the WIASRD data are informative, but using such data for statistical tests is not without participation bias. The WIASRD data offer only the information related to or reported by the participants. This situation results in possible participation bias of the data. It is therefore necessary to be cautious about the statistical inference based on these data.

Unit of Analysis

As noted above, the base unit of analysis for the study is individual program participants in a WIB[8] area in a state sorted by exit dates. Considering that the effects of WIA/WIOA programs are examined in this study, WIB areas are the natural base geographic level units, or the middle level of the study. Considering the fact that each state operates the federal government–funded WIA/WIOA programs in its own way, states are natural administrative geographic units that serve as the highest-level units for our hierarchical multilevel analysis. Workforce program participants might register in a local office in one county but find a job in another county, more likely than in another state. WIB areas are not necessarily labor market areas, residential areas, or legislative districts. Employment dynamics may occur across borders of WIB areas, but mostly remain within a state legislative boundary.

The Serial Autocorrelation Concern

The longitudinal WIASRD dataset makes serial autocorrelation a natural concern for the temporal dynamics affecting statistical estimates. Therefore, tests for serial correlation in the idiosyncratic errors of a linear panel-data model discussed by Wooldridge (2002) and Drukker (2003) are conducted. If serial autocorrelation were detected, the nature of the serial autocorrelation would be investigated to adopt appropriate temporal variables and model specific adjustments. In this study, no evident serial autocorrelation is detected. A reason is that the temporal unit is an older worker's exit date from the WIA/WIOA programs and services, which are not necessarily continuous, but discrete with gaps, in many cases. Also, a few days' exiting date difference does not necessarily result in an EER change. Therefore, the panel specific (by individual participant) serial autocorrelation concern was eliminated after the tests.

The Multilevel Mixed-Effects Regression Models

Three different types of regression models are used to test the hypotheses and test the robustness of the estimates. We started with simple logistic regression models estimated by the maximum likelihood methods, considering the binary nature of the dependent variables. Then considering the hierarchical nature of the data, we estimated mixed-effects regression models with linear probability estimate for the dependent variables. In the end, to integrate the advantage of logistic regression for binary dependent variables and the complexity of hierarchical nature of the data, a set of multilevel mixed-effects logistic regression models are estimated to capture the log of odds to be employed for our dependent variable (i.e., logit) and to control for the correlations at the WIB areas and at the policy structure level across states. The multilevel mixed-effects logistic regression models incorporates both fixed effects and random effects. We want to observe not only variations across specific individuals (fixed individual effects) but also variations across local WIB areas and across states (random WIB area and state effects). We want to observe both fixed and random effects on the binomial entered employment propensity (i.e., EER, as discussed above) across individuals who are living in different WIB areas or states—thus, our choice of multilevel mixed-effects logistic regressions.

Multilevel mixed-effects logistic regressions have been used extensively in various social science studies, such as Ng, Carpenter, Goldstein, and Rasbash (2006), which analyzed a Bangladeshi fertility survey; and Rabe-Hesketh and Skrondal (2012), which analyzed school data from Scotland. Rabe-Hesketh, Skrondal, and Pickles (2005) provide an excellent econometric survey on multilevel models with binary outcomes. As StataCorp (2015) notes, log-likelihood calculations for fitting any generalized mixed-effects model require integrating out the random effects.

The following equation briefly exhibits the multilevel mixed-effects model. Please note that the left hand side of the model is measuring older workers' EER. The right side of the equation measures program attributes, unemployment rates, and demographic-socioeconomic-program participation attributes:

$$\mathrm{EER}_{itws} = \beta_0 + \beta_1 \mathrm{P}_{itws} + \beta_2 \mathrm{U}_{itws} + \gamma \mathrm{DS}_{itws} + \mathrm{f(t\text{-}k)} + \varepsilon_{itws} + \mu_{itws} + \delta_{itws} + \zeta_{itws},$$

where

EER measures the entered employment probability, or log of odds in the logistic regression models, of entered employment (EE, a binary variable) during the first 3 months after an individual participant exited the program on the exiting date *t*.

P represents the WIA/WIOA program characteristics, including the aforementioned program types, training types, or occupational categories.

U is the unemployment rate level measure.

DS represents participants' individual characteristics, including aforementioned demographic, socioeconomic, and other workers' individual background information.

f(t) is a vector to capture the time effect, if there is any. In this case, it includes all the program year dummy variables to capture the time-varying unobserved factors.

Subscripts *i, t, w,* and *s* respectively specify individual workers, the exiting date, WIB areas, and state.

β_0, β_1, β_2, γ, and f are the regression parameters.

ε, μ, δ, and ζ are the error terms.

Dislocated Workers vs. Adults

The WIA/WIOA program have two different target population groups: *adults* and *dislocated workers*. Compared to dislocated workers, adults include all adults who are age 18 and above; the adult program is targeted to enhance adult workers' entered employment rate, employment retention, earnings and occupational skills, quality of the workforce, and reduced dependency on government-funded welfare and assistance programs. The priority for intensive and training services are given to low-income individuals and individuals receiving government assistance and with limited resources (ETA 2017d).

Dislocated workers instead refer to individuals who have been or to be terminated or laid off (particularly for business closures or structural unemployment reasons), who are eligible for or have exhausted unemployment insurance, who are unlikely to return to a previous industry or occupation, or who are displaced homemakers who are no longer supported by other family members (ETA 2017d). For them, job placement assistance and skill assessments and counseling could be helpful.

Dislocated workers are often better attached to the labor market than adults and could differ from adult workers in many perspectives, such as wage levels, skill levels, and prior attachment to the labor market. The study therefore model them separately. Two separate and independent datasets with unrelated attributes between dislocated worker participants and adult participants in the WIA/WIOA programs are drawn. They are estimated separately.[9]

Older Workers' Workforce Program Participation: Descriptive Statistics

This section describes characteristics of the observed workers. Figure 1 (next page) first exhibits older workers' overall demographic composition by age in the WIA/WIOA program. According to the WIASRD data, among those older worker participants who exited WIA/WIOA programs in 2011 through 2016, there were more older adult participants than older dislocated worker participants.[10] As a result, this dataset also shows that the number of older adults' entered employment is higher than that of older dislocated workers, though it does not mean older adults have a higher EER than older dislocated workers. In fact, as Table 2 (next page) shows, the EER among older dislocated workers is slightly higher (55%) than that among older adults (54%).

As *age* increases, program participation sharply decreases. There could be several reasons. First, as age increases and health conditions decline, older workers are less likely to participate in the labor market and less likely to seek training for a potential future job or job change. BLS (2017c) shows clearly that older individuals' labor force participation decreases with age. Second, as age increases, accessibility becomes a rising concern for some seniors to physically access local Workforce Investment Board offices, one-stop centers, and other facilities for training. Third, the WIA/WIOA program was in place for only about 15 years at the time the data were collected. This could result in limited awareness of training options among older workers. This limited awareness could be even more limited for older workers as age increases.

Table 2 exhibits the summary statistics for key variables we observe, with more details in the Appendix Table A1. The older workers in this study has a mean age at 57, are mostly (53%) male and white (68%), with limited representation in disabilities (7%), veterans (13%), offenders (4%), or limited English skills (1%). Poverty is an issue among those older workers. Despite limited representation in the TANF program (1%), homeless (2%), or receiving SSI or SSDI (2%), 16% received other public assistance, and 30% were low-income individuals.

FIGURE 1
Older WIA/WIOA Program Exiters, Adults vs. Dislocated Workers Overview by Age

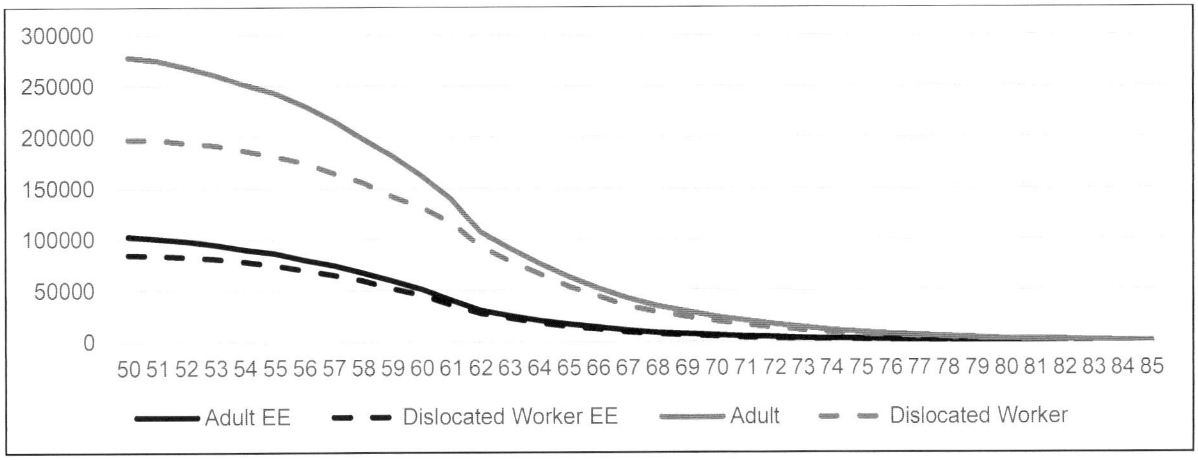

Data Source: WIASRD 2013-2015 Data

TABLE 2
Selected Participants and Participation Attributes

Variable	Obs	Mean	Std. Dev.	Min	Max
Adult EER	1,537,063	**0.5397**	0.50	0	1
Disl. Worker EER	1,272,513	**0.5540**	0.50	0	1
Adult	3,700,241	0.7078	0.45	0	1
Dislocated worker	3,700,241	0.5419	0.50	0	1
Demographics					
age	3,700,096	57.08	5.80	50	99
male	3,658,684	0.5298	0.50	0	1
Socioeconomic Status					
disabled	3,364,060	0.0663	0.25	0	1
veteran	3,699,555	0.1263	0.33	0	1
low income	3,001,450	0.3027	0.46	0	1
TANF	702,211	0.0089	0.09	0	1
SSDI	2,954,313	0.0207	0.14	0	1
other Public Assis	715,082	0.1582	0.36	0	1
homeless	706,194	0.0228	0.15	0	1
offender	718,534	0.0440	0.21	0	1
limited English	3,060,640	0.0119	0.11	0	1
Program Participation Information					
Wagner-Peyser Act	3,700,241	0.9139	0.28	0	1
Veterans program	393,025	0.3937	0.49	0	1
Received training	3,700,241	0.0977	0.30	0	1
Supportive scvs	3,700,241	0.0525	0.22	0	1
participation length	3,700,241	175.07	267	0	5481

For the program participation status, unsurprisingly, almost all participants (91%) received financial assistance services under the Wagner-Peyser Act., but only 10% received training and 5% received supportive services. While 39% received financial assistance services by both the LVER[11] Program and DVOP,[12] few received financial assistance services from other programs.

The mean program participation duration[13] is 175 days, but the standard deviation is almost 9 months. The longest participation duration last for 15 years. Many occurrence of long training lengths were from the youngest cohorts among senior WIA/WIOA participants.

Figure 2 shows education and credential information for the older workers. Almost half of them (43%) have gained occupational skills through certificates and 8% through licensure. However, only 5% reported having an associate's degree, 1% a bachelor's degree, and less than 1% had other educational attainment levels (high school or post-graduate degrees).

FIGURE 2
Education Attainments and Credentials

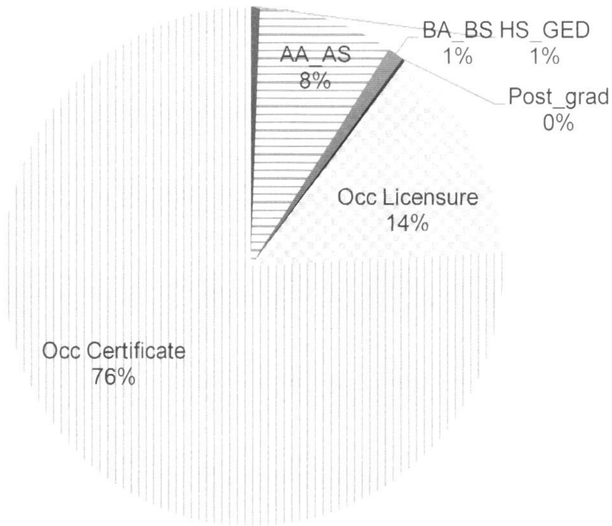

Data Source: WIASRD 2013-2015 Data

For the program participation, Figure 3 (next page) shows most older workers (67%) received other occupational skill training, 16% for skill upgrade training, 11% for on-the-job training, 5% accepted customized training, and few received adult and basic education or English as a second language training, or Entrepreneurial Training. For the occupational categories of training (Figure 4, next page), over 1/3 (38%) accepted management, administrative, professional and technical job skill training, 29% for mechanical transportation and military job trainings, 17% were in sales, clerk and administrative support job skill trainings, 13% ere in service work trainings, and 3% in agricultural, construction and extraction job trainings.

Older Adults vs. Older Dislocated Workers

As expected, dislocated workers are typically more attached to the labor market, has better and more knowledge-based skills, and therefore better off. Table 3 (two pages hence) exhibits the summary statistics for the t-tests, with more details provided in the appendix (Table A2). Compared to older adults, older dislocated workers were slightly older, with a slightly lower proportion of males and higher education attainment and occupation certificates but slightly lower occupation licensure; were more likely to be Hispanic, non-Hispanic Asian, or Caucasian; less likely to be non-Hispanic African American, Hawaiian, Pacific islanders, or American Indian; and less likely to be disabled, homeless, offenders, veterans, low-income, or receive SSDI or other public assistance, but slightly more likely to have limited English and a longer average program participation duration.

In terms of program participation attributes, compared to older adults, older dislocated workers are more likely to receive financial assistance services under the Wagner-Peyser Act, more likely to receive supportive services or trainings, more likely to receive services from veterans programs, TAA or NEG, and less likely to receive services from SNAP E&T and Pell grant.

FIGURE 3
Proportions Participating in Different Training Types

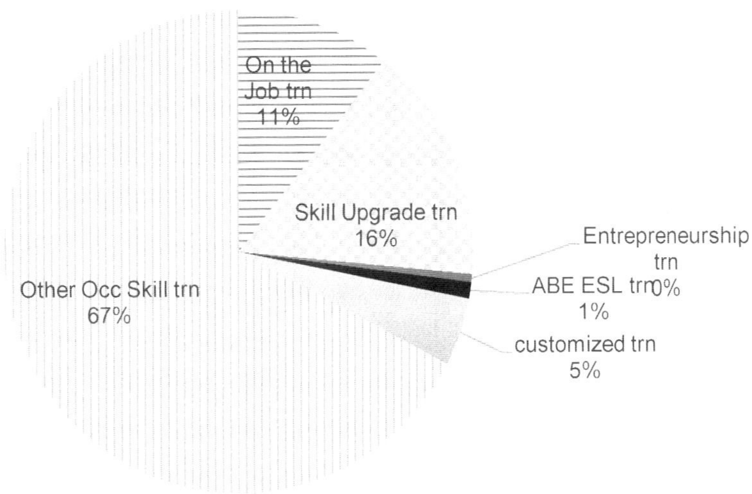

Data Source: WIASRD 2013-2015 Data

FIGURE 4
Proportion of Different Occupational Categories of Training

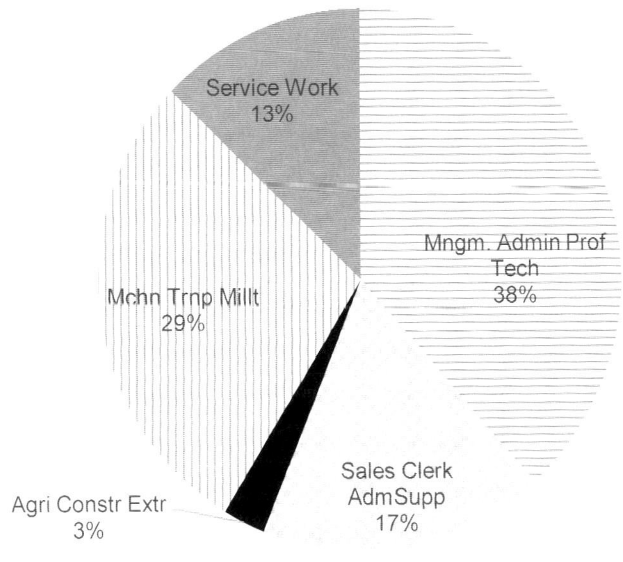

Data Source: WIASRD 2013-2015 Data

Compared to older adults, older dislocated workers tend to have better skills and participate in more high skill concentrated or knowledge-based occupation trainings. As Table 3 (next page) shows, older dislocated workers are also more likely to participate in management, administrative, professional, technical, or sales, clerk, and administrative support job trainings and less likely to participate in agricultural, construction, extraction, mechanical, transportation, military, and service work trainings.

TABLE 3

Participants and Participation Attributes Comparison, Older Dislocated Workers vs. Older Adults

	Dislocated Worker			Adult			
Variable	Obs	Mean	SD	Obs	Mean	SD	Sig
Demographics							
age	2,005,297	57.3	5.86	2,618,857	57.0	5.77	***
male	1,983,535	0.517	0.50	2,597,596	0.540	0.50	***
Socioeconomic Status							
Education attainment							
HS_GED	241,241	0.004	0.07	156,573	0.003	0.06	insig.
AA_AS	241,241	0.059	0.24	156,573	0.035	0.19	***
Post_grad	241,241	0.002	0.05	156,573	0.001	0.04	***
Occ Licensure	241,241	0.071	0.26	156,573	0.081	0.27	***
Occ Certificate	241,241	0.419	0.49	156,573	0.399	0.49	***
Race/Ethnicity							
Hispanic	1,892,186	0.095	0.29	2,501,712	0.079	0.27	***
Non-Hispanic Asian	1,892,186	0.034	0.18	2,501,712	0.021	0.14	***
Non-Hispanic African American	1,892,186	0.131	0.34	2,501,712	0.167	0.37	***
Non-Hispanic American Indian	1,892,186	0.007	0.08	2,501,712	0.012	0.11	***
Non-Hispanic White	1,892,186	0.712	0.45	2,501,712	0.698	0.46	***
disabled	1,778,818	0.049	0.22	2,363,156	0.076	0.26	***
veteran	2,004,878	0.116	0.32	2,618,742	0.140	0.35	***
low income	1,634,797	0.275	0.45	2,285,271	0.348	0.48	***
SSDI	1,575,988	0.013	0.11	2,242,502	0.025	0.16	***
other Public Assis	167,391	0.111	0.31	702,120	0.161	0.37	*
limited English	1,697,145	0.013	0.11	2,242,708	0.010	0.10	***
participation duration	2,005,327	205	291	2,618,993	146	239	***
Program Participation Information							
Wagner-Peyser Act	2,005,327	0.924	0.27	2,618,993	0.918	0.27	***
SNAP E&T	1,787,404	0.001	0.03	2,168,917	0.003	0.06	***
Pell Grant	244,973	0.037	0.19	157,542	0.045	0.21	**
Supportive scvs	2,005,327	0.059	0.23	2,618,993	0.036	0.19	***
Received training	2,005,327	0.122	0.33	2,618,993	0.060	0.24	***
Veterans program	177,967	0.423	0.49	297,089	0.352	0.48	***
TAA_NAFTA	1,922,584	0.032	0.18	2,448,283	0.007	0.09	***
NEG	2,005,327	0.004	0.07	2,618,993	0.001	0.03	***
Training Type							
On-the-Job trn	2,005,327	0.013	0.11	2,618,993	0.007	0.08	***
Skill Upgrade trn	2,005,327	0.020	0.14	2,618,993	0.009	0.10	***
Entrepreneurship trn	2,005,327	0.001	0.03	2,618,993	0.000	0.02	***
customized trn	2,005,327	0.001	0.04	2,618,993	0.005	0.07	***
Other Occ Skill trn	2,005,327	0.085	0.28	2,618,993	0.037	0.19	***
Occupational Categories of Training							
Mngm. Admin Prof Tech	216,815	0.410	0.49	138,181	0.336	0.47	***
Sales Clerk AdmSupp	216,815	0.189	0.39	138,181	0.160	0.37	***
Agri Constr Extr	216,815	0.025	0.16	138,181	0.030	0.17	***
Mchn Trnp Millt	216,815	0.265	0.44	138,181	0.323	0.47	***
Service Work	216,815	0.110	0.31	138,181	0.152	0.36	***

Empirical Findings: Regression Model Estimates

This section focuses on testing which factors affect older workers' entered employment (EE) probability after exiting the WIA/WIOA programs. When aggregating EE probability for a group of exiters, it becomes EER. We therefore use EER in the rest of the text to simplify the description. Table 4 (next page) displays the model estimates for older dislocated workers, and Table 5 (two pages hence) for older adults.[14] Each table presents estimates for the three different types of models: simple logistic regression model, mixed-effects logistic regression model, and multilevel mixed-effects logistic regression model. Please note that the coefficients presented for the former two model types are odds ratios, differing from the last model. Overall, across both sets of models, our estimates are highly similar and consistent. This reflects statistical rigor.

As shown in Table 4, overall the three models testing on older dislocated workers show consistent estimates. After dropping observation with missing values across all variables, only around 860 observations retained. Although the log-likelihood ratio test for multilevel models did not show statistical difference for the estimates compared to the simple logistic regression models, the variance controlled at the WIB Aarea levels shows statistical significance. The pseudo r-squared showed that the simple logistic regression estimates roughly explain 16% variability in older dislocated workers' EER.

Our findings in Table 4 supports our Hypothesis 1 among older dislocated workers. For program participation attributes, receiving financial assistance services under the Wagner-Peyser Act reduced EER, and this is consistent across all three models. The Wagner-Peyser Act of 1933, known as the Employment Service, established a nationwide system of public employment offices and went through amendments in 1998 as part of the one-stop delivery system under the WIA and in 2014 to align performance accountability indicators with other federal workforce programs under the WIOA (ETA 2017e). Employment services provided with Wagner-Peyser Act funding are available to all job seekers and employers (O'Leary and Ebert 2008). Many of the participants under Wagner-Peyser use either self-assisted services or staff-assisted services. This limited intensity of service, compared to intensive training, is unsurprisingly associated with lower EER.

Supportive service contributed to higher EER, according to the logistic regression and the multilevel mixed-effects model, though with limited statistical significance ($p=0.1$). The supportive services include, but are not limited to, assistance with transportation, child care, dependent care, and housing that are necessary to enable the individual to participate in activities authorized under WIA/WIOA Title IB. For those older workers who have accessibility issues or with responsibility for children, this could be helpful.

For different training types, on-the-job training, skill upgrade training, and adult basic education and English as a second language training are shown to be highly effective to enhance EER among older dislocated workers. This is particularly so for on-the-job training. Participating in on-the-job training increased the EER by 22 folds according to the mixed-effects logistic regression model and by 19 folds according to the simple logistic regression model, by 54% according to the multilevel mixed-effects regression model, holding all other variables constant. On-the-job training connects workers directly with job skills needed in the market and may also equip trainees with the necessary network. This could greatly enhance job placement odds. The skill upgrade training, and adult basic education and English as a second language training are also highly effective, though the effects are not as strong as the on-the-job training. Since dislocated workers are often structurally unemployed workers, skill upgrading, updating, and retraining could be beneficial to find a new job.

To test the Hypothesis 2 among older dislocated workers, the monthly state unemployment rate did not has statistically significant effect (at $p= 0.1$) on EER. Compared to adults or other participants, dislocated workers are relatively better off and therefore are less vulnerable to business cycle changes. The study focus on the post-recession economic recovery period. The slight downturns during the recovery might not have a strong enough consistent impact on dislocated workers. However, this effect could be more sensitive among older adults.

TABLE 4
Regression Estimates for Older Dislocated Workers

	Mixed-Effects Logistic Regression		Simple Logistic Regression		Multilevel Mixed-Effects Regression	
	Odds Ratio		Odds Ratio		Coef.	
Unemployment Rate	1.02		1.00		0.00	
Demographics						
age	0.92	***	0.92	***	-0.02	***
male	0.70		0.74		-0.07	
Race/Ethnicity						
Asian_nonH	0.09	*	0.10	*	-0.50	**
American Indian nonH	0.11		0.10		-0.39	*
Other race/ethnic variables that are insignificant: Hispanic, non-Hispanic African Americans, non-Hispanic Hawaiian Pacific Islanders, and non-Hispanic White.						
Socioeconomic Status						
Education attainment						
HS_GED	4.50		4.38		0.29	
AA_AS	0.41	**	0.46	**	-0.17	***
BA_BS	1.71		1.43		0.10	
Occ_Licensure	1.71	*	1.73	*	0.09	*
Occ _Certificate	1.02		1.02		0.00	
homeless	0.27	*	0.30	*	-0.23	*
limited English	1.00		1.00		0.41	*
Other socioeconomic variables that have insignificant effects: disabled, veteran, low income, TANF, SSDI, Other_PubAssis, offender, participation duration.						
Program Participation Information						
Wagner-Peyser Act	0.21	***	0.21	***	-0.22	***
Supportive svcs	1.45		1.49	*	0.07	*
Needs-Related Payments	1.00		1.00		0.00	
Other program participation information with insignificant effects: Pell Grant, distance learning, Received training, Veterans program, TAA/NAFTA, ARRA, apprentice training, other basic skill Training, NEG, and Older American Act 1998.						
Training Type						
On-the-Job training	22.65	***	20.23	***	0.54	***
Skill Upgrading training	9.17	**	8.34	**	0.47	**
Entrepreneurship training	1.00		1.00		0.00	
ABE ESL training	19.28	*	18.38	*	0.58	**
Customized training	4.50		4.16		0.33	
Other Occ Skill training	4.77		4.52		0.33	*
Occupational categories of training: Insignificant occupational categories of training: Management Admin Professional Technician, Sales Clerk Admin support, Service Work, Mechanics Transportation Military						
	Year dummies for year 2011 (***, +), 2012, 2013 (***, -), 2014, 2015					
Constant	929	***	1039	***	2	***
State						
var(_cons) or SD	4E-37	.			1E-10	
# of groups, # obs per group	8,	107			8,	109
State>WIB area						
var(_cons) or SD	0.08	**			0.05	**
# of groups, # obs per group	91,	9			91,	10
sd(Residual)					0.42	**
LR test vs. logistic (or linear) model	0.7				0.74	
# of Obs	856		856		863	
Integration method, Integration Point	Mvaghermite,	7				
Wald chi2(47)	119.09	***			196.35	***
Log likelihood =	-451		-451		-477	
LR chi2(48)			173	***		
Pseudo R2			0.16			

TABLE 5
Regression Estimates for Older Adults

EER	Mixed-Effects Logistic Regression		Simple Logistic Regression		Multilevel Mixed-Effects Regression	
	Odds Ratio		Odds Ratio		Coef.	
Unemployment Rate	0.81	**	0.78	***	-0.04	**
Demographics						
age	0.93	***	0.94	***	-0.01	***
male	0.85		0.84		-0.03	
Race/Ethnicity						
Non-Hispanic Asian	0.14	**	0.16	*	-0.41	**
Other race/ethnic variables that are insignificant: Hispanic, Non-Hispanic African American, Non-Hispanic Hawaiian Pacific Islander., Non-Hispanic American Indian, Non-Hispanic White						
Socioeconomic Status						
Education attainment						
HS_GED	1.00		1.00		0.43	*
AA_AS	0.86		0.89		-0.03	
BA_BS	2.08		1.84		0.16	
Occ_Licensure	2.20	***	2.33	***	0.14	***
Occ_Certificate	1.15		1.21	*	0.03	
disabled	0.71	**	0.71	**	-0.07	**
veteran	2.27		2.69	*	0.15	
TANF	2.94		2.97		0.20	*
Other Public Assistance	0.72	**	0.71	**	-0.07	**
homeless	0.61	*	0.58	**	-0.10	**
participation duration	0.9995	***	0.9995	***	-0.0001	***
Other socioeconomic variables that have insignificant effects: SSDI, low income, offender, limited English						
Program Participation Information						
Needs-Related Payments	0.05	**	0.05	**	-0.56	***
Other program participation information with insignificant effects: Pell Grant, distance learning, Received training, Veterans program, TAA/NAFTA, ARRA, apprentice training, other basic skill Training, NEG, Wagner-Peyser Act, Older American Act 1998, Supportive svcs.						
Training Type						
On-the-Job training	13.59	***	9.53	**	0.49	***
Other insignificant training type variables: Skill Upgrading training, Entrepreneurship training, ABE ESL training, customized train, Other Occupational Skill training						
Occupational Categories of Training						
Training Mechanic Transportation Military	1.54	*	1.53	*	0.08	*
Other insignificant occupational categories of training: Management Admin Professional Technician, Sales Clerk Admin Support, Service Work						
	Year dummies for year 2011 (***, +), 2012, 2013 (***, -), 2014, 2015					
Constant	42	**	58	***	1	***
State						
var(_cons) or SD	2E-03	**			3E-06	**
# of groups, # obs per group	9,	264			9,	265
State>WIB area						
var(_cons) or SD	0.13				0.07	**
# of groups, # obs per group	145,	16			145,	17
sd(Residual)					0.44	**
LR test vs. logistic (or linear) model:	17.08	***			17.7	***
# of Obs	2380		2380		2387	
Integration method, Integration Point	Mvaghermite,	7				
Wald chi2(47)	226.88	***			302.06	***
Log likelihood =	-1361		-1369		-1434	
LR chi2(48)			301	***		
Pseudo R2			0.10			

For demographic control variables, older ages are associated with a lower entered employment (EE) probability, controlling for all other variables. In the mixed-effects logistic regression and simple logistic regression, both show that being one year older reduced the odds of being employed during the first quarter after exiting the program by 8%; consistently, in the multilevel mixed-effects regression model, older age is related to a lower EE probability. Gender does not display an effect on older dislocated workers' EER. Compared to mixed races, being non-Hispanic Asian reduced the EER; being non-Hispanic American Indian reduced the EER only for the multilevel mixed-effect regression mode (at $p=0.1$). Zhang (2011) found that African American older workers were the most vulnerable group in terms of EER, using WIASRD 2007 data; however, this is not the case anymore. This could be related to the recent years' effective policy implementation to empower African Americans. Although Asians' limited EER could be associated with a longer search period for better-fit jobs or could be associated with cultural and language adaptation if they were immigrants, certain policy attention and investigation of Asian older workers could be useful.

The socioeconomic control variable effects are consistent across all the three models. Education and credentials are important to employment. While having attained an associate's degree reduced the EE odds, having occupational licensure increased the EE odds. Homeless older dislocated workers have lower EER.

Table 5 presents the estimates for older adults. Overall, the three models testing on older dislocated workers again show consistent and highly similar estimates. About 2380 or 2387 observations are retain for the model estimates across all variables. The log-likelihood ratio test for the multilevel model estimates this time showed clear statistical difference, compared to the simple logistic regression models; the variance controlled at both the state and the WIB area levels also showed statistical significance. Therefore, multilevel modeling display advantage over simple logistic model among older adults. The pseudo r-squared showed that the simple logistic regression estimates roughly explain 10% variability in older dislocated workers' EER.

Our Hypothesis 1 is again supported among older adults, though with the mechanism somewhat different from that among older dislocated workers. Consistent with the findings among older dislocated workers, on-the-job training is highly effective in enhancing older adults' EER as well. Participating in on-the-job training enhances older adults' EER by 9 or 13 folds, according to respectively the simple logistic regression and the mixed-effects logistic regression models, or by 49%, according to the multilevel mixed-effects regression estimate. On-the-job training connects skills better with job contents, markets, and network, which could benefit job hunting. Differing from the findings among older dislocated workers, none of the other training types matter to the older adults' EER. While skill upgrading could help dislocated workers who typically have better skills to be upgraded upon, skill upgrading does not seem work well among older adults.

Although none of the occupation categories of training matters to older dislocated workers, participating in mechanical, transportation, and military skill training enhances older adults' EER. Older adults are less driven by the knowledge-based economic skills, as our descriptive statistics showed earlier; however, training them with skills that are more physical could help them place a job.

While receiving need-related payments does not matter to older dislocated workers, older adults receiving need-related payments have lower EER. This is a particularly vulnerable subgroup among older adults and particularly need assistance in financial, human, and social capitals.

Among older adults, those who participated in the program for a longer time period tend to have a reduced EER. This effect is highly significant (at $p<0.01$). The magnitude seems very small—for the odds ratio; it is almost 1, which means no effect; for the multilevel mixed-effect regression, it is close to 0. However, the magnitude is relatively to a day's change. If changing the participation duration scale to a month or a year, the magnitude would become much larger.

Among older adults, the cyclical effects in our Hypothesis 2 is evident, different from the aforementioned findings among older dislocated workers. Older adults' EER were impacted by the monthly state unemployment rate, controlling for other variables. As mentioned earlier, adults are socioeconomically worse off than dislocated workers, less attached to the labor market, and therefore are typically more vulnerable to labor market changes.

For the demographic control variable effects, the findings were the same among older dislocated workers: older age and being non-Hispanic Asians again are both associated with lower EER among older adults. For education and credentials, compared to older dislocated workers, academic degrees do not matter as much among older adults. As indicated earlier, older adults are typically less-educated than older dislocated workers. Among older adults, having a high school or equivalent degree seems to be associated with better EER only in the multilevel mixed-effects regression model with limited statistical significance (at p=0.1), but this is insignificant for mixed-effect logistic regression or simple logistic regression models. However, occupation licensure again shows a strong positive effect in enhancing EER. Having occupational licensure increases the EER by 1.2 to 1.3 fold, according to the simple and mixed-effects logistic regression models and increase the probability by 14% according to the multilevel mixed-effect regression model. Among older adults, occupational certificate also display a potential positive effect on EER in the simple logistic regression, though with limited statistical significance (p=0.1).

Since adults are socioeconomically worse off than dislocated workers, the most vulnerable include the disabled, veterans, TANF recipients, the homeless, and those receiving other public assistance programs and receiving needs-related payments. They therefore had a lower EER. Those vulnerable subpopulation groups deserve more policy support.

Conclusion

This paper relies on the WIASRD 2013-2015 data and BLS unemployment data and adopts three types of logistic regression models to identify factors associated with older workers' EER after exiting WIA/WIOA programs. The three models are respectively simple logistic regression, mixed-effects regression, and multilevel mixed-effects logistic regression models. The three model types are estimated separately for older dislocated workers and older adults.

Main Findings

Findings across the simple logistic regression, mixed-effects regression, and multilevel mixed-effects logistic regression models are highly similar and consistent. This demonstrates statistical rigor of our data and modeling. Multilevel mixed-effects modeling shows some advantage in either log-likelihood ratio tests or in controlling variance at a higher analysis level (either at the WIB area level or at the state level). Although pseudo r-squared is not super high, it stays at 10% or 16% for the simple logistic regression; for an individual-level study with a high level of unobserved heterogeneity across individual participants, this is acceptable.

Our findings support our first hypothesis among both older dislocated workers and older adults, but support our second hypothesis only among older adults. For the findings on Hypothesis 1, on-the-job training is particularly effective for a better EER. Participating in the on-the-job training greatly increases an older participant's EER by up to 22 folds. On-the-job training directly connect trainees to job skills in demand and related social networkd, which turns out to be highly effective. Our models also show that supportive services, skill upgrade training, and adult basic education and English as a second language training are not helpful to older adults, but they are effective in enhancing older dislocated workers' EER. Older dislocated workers are socioeconomically better off, and they have different needs from older adults. With adequate skills for jobs, offering supportive services, such as transportation assistance or childcare, or even care for other family members, could relieve older dislocated workers from barriers to go to work. Skill upgrade training helps dislocated worker who are typically structurally unemployed to upgrade and update their skills. Adult basic education and English training could help older immigrants or older workers who have strong learning desire and capability to advance well. Those programs, located often either at workforce training centers or community colleges, could be relatively small but effective.

Since older adults are more likely to concentrate on less knowledge-based jobs, targeting funds for their needs should orient them more toward the skills they have historical background in. For example, our models shows that participating in mechanical, transportation, and military skill trainings evidently help older adults to have a higher EER. Training funds for older adults can be more oriented toward those occupational categories of training.

For the program participation duration, although older dislocated workers have a longer average participation duration, long participation duration among dislocated workers did not affect their EER. However, long participation duration among older adult workers affected their EER. This implies the importance of monitoring participation duration particularly among older adults to limit repeating services and redundant programs to the same individuals. Also, follow-up and monitoring using Big Data tools can help identify the people who have multiple spells and whether the multiple spells are for the different services and whether that is necessary.

For findings on Hypothesis 2, our models show that a higher unemployment rate reduces older adults EER, while older dislocated workers are not sensitive to the monthly state unemployment rate change. Older adults are particularly vulnerable in the post-recession labor market cyclical changes. Older adults are more socioeconomically vulnerable and therefore could be more affected in an economic downturn. They are also more likely than older dislocated workers to have mobility limitation. Local state labor market conditions could be more important to them. Therefore, equipping older adults with better job information flow could help expose them to more job opportunities and make them less vulnerable to local job conditions.

Policy Implications

The empirical tests identify policy implications. The findings suggest a need to draw policy attentions in several specific directions. For older adults, on-the-job training and trainings on mechanical, transportation and military skills can help them most effectively. Following up and monitoring older adults program participation duration using Big Data or other methods and try to reduce unnecessary redundant services could help enhance the cost effectiveness of the program.

For older dislocated workers, on-the-job training, skill upgrade training, adult basic education, and English as a second language training are the most effective training methods. Focusing on supportive services is important to help remove their barriers to employment.

Consistent with expectations, our data show that older adults are socioeconomically more vulnerable than older dislocated workers. Particularly older adults who are disabled, veterans, recipients of TANF, needs-related payments, or other public assistance are the one with a lower EER; they need more policy support. Those who are older, homeless and non-Hispanic Asian were found in this study with lower EER and therefore deserve policy attention for better assistance. Those who with occupational licensure or participated in on-the-job training were found to have a higher EER. Occupation oriented licensure are very effective in the job market among older Americans.

Implications on Training Methods

Our literature review suggests implications on training methods for older workers. Due to path dependency, trainings based on previous job experience and skills could be more effective. With older workers' decreasing vision and hearing, high contrast settings for print or computer screens or audio setting can help stimulate visual and audio learning. Also as Fisk et al. (2004) suggested, matching the instructional technique and medium would help; performing more action based hand-on procedural training, instead of conceptual could not only be more interesting to trainees, but useful to enhance training effectiveness for older trainees. Allowing extra time, such as 1.5 to 2 times of the training time expected for younger adults, to process information is suggested by Charness and Czaja (2006). Also, considering the fact that many older workers have not looked for a job in recent years, job search and interview skills training would be useful as well.

Future Research Directions

Further exploration on other potential factors, such as industry details, more occupational details, and spatial effects, could be helpful to add more nuances. Different industry and occupational setting may require different training mechanisms to be the most effective. An analysis detailed down to industry and occupational level may reveal important implications.

While the Web technically advances fast, often the hands-on trainings only occur locally. Also, the local labor market might have a spatial spillover effect that results in a certain level of spatial autocorrelation. People search jobs across states or metropolitan areas. Therefore,a follow-up study with focus on the spatial effects and with spatial controls could add value to understand what helps make workforce programs better.

Exploring temporal changes across the years, particularly after the WIOA started, could be another interesting extension. This study captures the beginning period of WIOA era, but is mostly still under WIA. In a few years when data on WIOA program participation are available, an investigation on the policy change from WIA to WIOA could be another avenue of value-added research for further policy and practice implications.

Appendix

TABLE A1

Description of Variables for Older WIA/WIOA Program Participants Who Are Dislocated Workers

Variable	Obs	Mean	Std. Dev.	Min	Max
Adult EER	1,537,063	**0.54**	0.50	0	1
Disl. Worker EER	1,272,513	**0.55**	0.50	0	1
adult	3,700,241	0.71	0.45	0	1
Dislocated worker	3,700,241	0.54	0.50	0	1
Demographics					
age	3,700,096	57.08	5.80	50	99
male	3,658,684	0.53	0.50	0	1
Socioeconomic Status					
Education attainment					
HS/GED	357,022	0.00	0.07	0	8
AA/AS	357,022	**0.05**	0.22	0	8
BA/BS	357,022	**0.01**	0.08	0	8
Post-grad	357,022	0.00	0.05	0	8
Occ Licensure	357,022	**0.08**	0.27	0	8
Occ Certificate	357,022	**0.43**	0.50	0	8
Race/Ethnicity					
Hispanic	3,505,214	0.09	0.29	0	1
Non-Hispanic Asian	3,505,214	0.03	0.17	0	1
Non-Hispanic Black	3,505,214	0.17	0.37	0	1
Non-Hispanic Hawaiian & Pacific Islander	3,505,214	0.00	0.05	0	1
Non-Hispanic American Indian	3,505,214	0.01	0.10	0	1
Non-Hispanic White	3,505,214	0.68	0.47	0	1
disabled	3,364,060	0.07	0.25	0	1
veteran	3,699,555	0.13	0.33	0	1
low income	3,001,450	0.30	0.46	0	1
TANF	702,211	0.01	0.09	0	1
other Public Assis	715,082	0.16	0.36	0	1
homeless	706,194	0.02	0.15	0	1
offender	718,534	0.04	0.21	0	1
limited English	3,060,640	0.01	0.11	0	1
Program Participation Information					
vocational educ	3,015,535	0.00	0.01	0	1
Wagner-Peyser Act	3,700,241	0.91	0.28	0	1
Older American Act 1998	2,985,897	0.00	0.01	0	1
SNAP E&T	3,050,422	0.00	0.05	0	1
Pell Grant	361,357	0.04	0.19	0	1
distance learning	361,357	0.00	0.06	0	1
Supportive service	3,700,241	0.05	0.22	0	1
Need based service	1,276,196	0.00	0.06	0	1
Receiving Training	3,700,241	0.10	0.30	0	1
SSDI	2,954,313	0.02	0.14	0	1
Veterans program	393,025	0.39	0.49	0	1
TAA_NAFTA	3,491,786	0.02	0.14	0	1
Vocational Rehabilitation	2,995,448	0.00	0.02	0	1
ARRA	3,700,241	0.04	0.19	0	1
Other Program	3,700,241	0.01	0.09	0	1
apprentice training	3,700,241	0.00	0.00	0	1
other basic skill training	3,700,241	0.00	0.00	0	1
NEG	3,700,241	0.00	0.05	0	1

Table A1 continues, next page

Variable	Obs	Mean	Std. Dev.	Min	Max
Training Type					
On-the-Job Training	3,700,241	0.01	0.10	0	1
Skill Upgrade Training	3,700,241	0.02	0.12	0	1
Entrepreneurship Training	3,700,241	0.00	0.02	0	1
ABE ESL Training	3,700,241	0.00	0.03	0	1
customized Training	3,700,241	0.00	0.07	0	1
Other Occupational Skill Training	3,700,241	0.06	0.25	0	1
Occupational Categories of Training					
Management Admin Prof Tech	319,302	0.38	0.49	0	1
Sales Clerk Admin Support	319,302	0.18	0.38	0	1
Agricultural Construction Extraction	319,302	0.03	0.16	0	1
Mechanics Transportation Military	319,302	0.29	0.45	0	1
Service Work	319,302	0.13	0.33	0	1
Participation. length	3,700,241	175	267	0	5481

TABLE A2

Participants and Participation Attributes Comparison, Older Dislocated Workers vs. Older Adults

	Dislocated Worker			Adult			
Variable	Obs	Mean	Std. Dev.	Obs	Mean	Std. Dev.	Sig
Adult EER	568,555	0.552	0.50	1,537,063	0.540	0.50	
Disl. Worker EER	1,272,513	0.554	0.50	568,555	0.552	0.50	
Adult	2,005,327	0.461	0.50	2,618,993	1.000	0.00	
Dislocated worker	2,005,327	1.000	0.00	2,618,993	0.353	0.48	
Demographics							
age	2,005,297	57.3	5.86	2,618,857	57.0	5.77	***
male	1,983,535	0.517	0.50	2,597,596	0.540	0.50	***
Socioeconomic Status							
Education attainment							
HS_GED	241,241	0.004	0.07	156,573	0.003	0.06	insig.
AA_AS	241,241	0.059	0.24	156,573	0.035	0.19	***
BA_BS	241,241	0.007	0.09	156,573	0.004	0.07	insig.
Post_grad	241,241	0.002	0.05	156,573	0.001	0.04	***
Occ Licensure	241,241	0.071	0.26	156,573	0.081	0.27	***
Occ Certificate	241,241	0.419	0.49	156,573	0.399	0.49	***
Race/Ethnicity							
Hispanic	1,892,186	0.095	0.29	2,501,712	0.079	0.27	***
Non-Hispanic Asian	1,892,186	0.034	0.18	2,501,712	0.021	0.14	***
Non-Hispanic African American	1,892,186	0.131	0.34	2,501,712	0.167	0.37	***
Non-Hispanic Hawaiian Pac Isl.	1,892,186	0.002	0.05	2,501,712	0.003	0.05	insig.
Non-Hispanic American Indian	1,892,186	0.007	0.08	2,501,712	0.012	0.11	***
Non-Hispanic White	1,892,186	0.712	0.45	2,501,712	0.698	0.46	***
disabled	1,778,818	0.049	0.22	2,363,156	0.076	0.26	***
veteran	2,004,878	0.116	0.32	2,618,742	0.140	0.35	***
low income	1,634,797	0.275	0.45	2,285,271	0.348	0.48	***
TANF	166,324	0.006	0.07	689,460	0.009	0.09	insig.
SSDI	1,575,988	0.013	0.11	2,242,502	0.025	0.16	***
other Public Assis	167,391	0.111	0.31	702,120	0.161	0.37	*
homeless	169,624	0.015	0.12	690,250	0.023	0.15	insig.
offender	170,456	0.027	0.16	702,533	0.045	0.21	insig.
limited English	1,697,145	0.013	0.11	2,242,708	0.010	0.10	***
participation duration	2,005,327	205	291	2,618,993	146	239	***

Table A2 continues, next page

	Dislocated Worker			Adults			
Variable	Obs	Mean	Std. Dev.	Obs	Mean	Std. Dev.	Sig
Program Participation Information							
vocational educ	1,774,228	0.000	0.01	2,145,330	0.000	0.01	insig.
Wagner Pyser Act	2,005,327	0.924	0.27	2,618,993	0.918	0.27	***
Older American Act 1998	1,757,660	0.000	0.01	2,132,913	0.000	0.01	insig.
SNAP E&T	1,787,404	0.001	0.03	2,168,917	0.003	0.06	***
Pell Grant	244,973	0.037	0.19	157,542	0.045	0.21	**
distance learning	244,973	0.004	0.06	157,542	0.002	0.04	insig.
Supportive scvs	2,005,327	0.059	0.23	2,618,993	0.036	0.19	***
Needs-Related Payments	728,089	0.004	0.06	702,564	0.002	0.04	insig.
Received training	2,005,327	0.122	0.33	2,618,993	0.060	0.24	***
Veterans program	177,967	0.423	0.49	297,089	0.352	0.48	***
TAA_NAFTA	1,922,584	0.032	0.18	2,448,283	0.007	0.09	***
ARRA	2,005,327	0.036	0.19	2,618,993	0.039	0.19	insig.
apprentice training	2,005,327	0.000	0.00	2,618,993	0.000	0.00	insig.
other basic skill Training	2,005,327	0.000	0.00	2,618,993	0.000	0.00	insig.
NEG	2,005,327	0.004	0.07	2,618,993	0.001	0.03	***
Training Type							
On-the-Job trn	2,005,327	0.013	0.11	2,618,993	0.007	0.08	***
Skill Upgrade trn	2,005,327	0.020	0.14	2,618,993	0.009	0.10	***
Entrepreneurship trn	2,005,327	0.001	0.03	2,618,993	0.000	0.02	***
ABE ESL trn	2,005,327	0.001	0.03	2,618,993	0.001	0.03	insig.
customized trn	2,005,327	0.001	0.04	2,618,993	0.005	0.07	***
Other Occ Skill trn	2,005,327	0.085	0.28	2,618,993	0.037	0.19	***
Occupational Categories of Training							
Mngm. Admin Prof Tech	216,815	0.410	0.49	138,181	0.336	0.47	***
Sales Clerk AdmSupp	216,815	0.189	0.39	138,181	0.160	0.37	***
Agri Constr Extr	216,815	0.025	0.16	138,181	0.030	0.17	***
Mchn Trnp Millt	216,815	0.265	0.44	138,181	0.323	0.47	***
Service Work	216,815	0.110	0.31	138,181	0.152	0.36	***

References

Beier, M. E., and Ackerman, P. L. (2005). "Age, Ability, and the Role of Prior Knowledge on the Acquisition of New Domain Knowledge: Promising Results in a Real-World Learning Environment." *Psychology and Aging, 20,* 341-355.

Brown, S. K. (2003). "Staying Ahead of the Curve 2003: The AARP Working in Retirement Study." *AARP Knowledge Management*: Washington DC.

Bureau of Labor Statistics (BLS) (2017a). *Labor Force Statistics from the Current Population Survey*. http://bit.ly/2mOxfrt

Bureau of Labor Statistics (BLS) (2017b) *Employment Projections: Civilian labor force by detailed age, gender, race, and ethnicity*, http://bit.ly/2mMHwnW

Bureau of Labor Statistics (BLS) (2017c). *Employment Projections: Civilian labor force participation rate by age, gender, race, and ethnicity*. http://bit.ly/2mOKQiC

Carlson, M. C., Hasher, L., Zacks, R. T., and Connelly, S. L. (1995). "Aging, Distraction, and the Benefits of Predictable Location." *Psychology and Aging, 10*, 427-436.

Charness, N. (2004). "The Age-Ability-Productivity Paradox." Wharton Impact Conference Proceedings, *Maximizing Your Labor Force: Employees Over 50 In Today's Global Economy*. Nov. 10, 2004. University of Pennsylvania and AARP Global Aging Program.

Charness, N. and Czaja, S. J. (2006), "Older Worker Training: What We Know and Don't Know," AARP working paper #2006-22.

Charness, N., Schumann, C. E., and Boritz, G. A. (1992). "Training Older Adults in Word Processing: Effects of Age, Training Technique and Computer Anxiety." *International Journal of Aging and Technology, 5*, 79-106.

Collison, J. (2003). (program manager). "Older Workers Survey." SHRM research, June 2003. Society for Human Resource Management (SHRM), National Older Worker Career Center (NOWCC) and Committee for Economic Development (CED).

Colquitt, J. A., LePine, J. A., and Noe, R. A. (2000). "Toward an Integrative Theory of Training Motivation: A Meta-Analytic Path Analysis of 20 Years of Research." *Journal of Applied Psychology, 85, 678-707.*

Crown, W. H. and Longino, C. F., Jr. (2000). "Labor Force Trends and Aging Policy." E. W. Markson and L. A. Hollis-Sawyer (Eds.). *Intersections of aging: Readings in social gerontology*. Los Angeles: Roxbury.

Czaja, S. J., Hammond, K., Blascovich, J., and Swede, H. (1989). "Age-Related Differences in Learning to Use a Text-Editing System." *Behavior and Information Technology 8*, 309-319.

Drukker, D. M. 2003. Testing for serial correlation in linear panel-data models. Stata Journal (3)2: 168-177.

Elias, P. K., Elias, M. F., Robbins, M. A., and Gage, P. (1987). "Acquisition of Word- Processing Skills by Younger, Middle-Aged, and Older Adults." *Psychology and Aging, 2*, 340-348.*Psychology, 85, 678-707.*

Employment and Training Administration (ETA) (2017a). *WIOA Overview*. http://bit.ly/2Dqi2Ix

Employment and Training Administration (ETA) (2017b). *PY 2015 WIA SRD Data Book*, http://bit.ly/2DmQJi9

Employment and Training Administration (ETA) (2017c). *Senior Community Service Employment Program*. http://bit.ly/2DmhgvO

Employment and Training Administration (ETA) (2017d). *Workforce Investment Act -- Adults and Dislocated Workers Program*. http://bit.ly/2DrthjO

Employment and Training Administration (ETA) (2017e). *Wagner-Peyser Act Employment Service*. http://bit.ly/2DrtL9C

Eyster, L., Johnson, R. W., and Toder, E. (2008), "Current Strategies to Employ and Retain Older Workers," Urban Institute Report to the Department of Labor, Employment and Training Administration. http://urbn.is/2DmTqjD

Friedberg, L. (2003). "The Impact Of Technological Change On Older Workers: Evidence from Data on Computer Use," Industrial and Labor Relations Review, v56 (3, Apr), 511-529.

Fisk, A. D., Rogers, W. A., Charness, N., Czaja, S. J., and Sharit, J. (2004). *Designing for Older Adults: Principles and Creative Human Factors Approaches*. Boca Raton, Florida: CRC Press.

Gist, M., Rosen, B., and Schwoerer, C. (1988). "The Influence of Training Method and Trainee Age on the Acquisition of Computer Skills." *Personal Psychology, 41*, 255-265.

Heidkamp, M. (2009), "Older and Out of Work: Trends in Older Worker Displacement," SCSEP Senate Briefing, Rutgers University. http://bit.ly/2DpCpWb

Hooymann, N. R. and Kiyak, H. A. (2005). *Social Gerontology: A multidisciplinary Perspective*. Boston: Allyn and Bacon. Chapter 1, 8, 9, 12, 14, 16.

Human Resources Development Canada, (1999). "Older Worker Adjustment Programs Lessons Learned (Final Report): Evaluation and Data Development Strategic Policy." Human Resources Development Canada, Government of Canada: SP-AH093-12-99E. http://bit.ly/2Dp0axv

Kautonen T, Down S, and South L (2008) Enterprise support for boomer entrepreneurs: The case of PRIME in the UK. *International Journal of Entrepreneurial Behaviour and Research* 14(2): 85–101.

Office of Technology Assessment. Worker Training: Competing in the New International Economy. Washington, DC: OTA, September 1990. (ED 326 622).

Lavarreda, S. A., Snyder, S., and Brown, E. R. (2013). The effects of the great recession on health insurance: Changes in the uninsured population from 2007 to 2009. *UCLA Center for Health Policy Research, Health Policy Brief, July 2013.*

Maurer, T., Weiss, M, and Barbeite, F. (2003). "A Model of Involvement in Work-Related Learning and Development Activity: The Effects of Individual, Situational, Motivational and Age Variables." *Journal of Applied Psychology, 88*, 707-724.

Mead, S., and Fisk, A. D. (1998). "Measuring skill acquisition and retention with an ATM simulator: The need for age-specific training." *Human Factors, 40*, 516-523.

Mead, S. E., Spaulding, V. A., Sit, R. A., Meyer, B., and Walker, N. (1997). "Effects of Age and Training on World Wide Web Navigation Strategies." *Proceedings of the Human Factors and Ergonomics Society 41st Annual Meeting*, 152-156. Santa Monica, California: Human Factors and Ergonomics Society.

Morrell, R. W., Park, D. C., Mayhorn, C. B., and Echt, K. V. (1995). *Older Adults and Electronic Communication Networks: Learning to Use ELDERCOMM.* Paper presented at the 103rd Annual Convention of the American Psychological Association. New York.

Munnell, A., Muldoon D., and Sass, S. A., (2009). *Recessions & Older Workers*, IB#9-2, Chestnut Hill, MA: Center for Retirement Research at Boston College, 10 pp. http://crr.bc.edu.

Ng, E. S.-W., Carpenter, J. R., Goldstein, H., and Rasbash, J. (2006). Estimation in generalized linear mixed models with binary outcomes by simulated maximum likelihood. *Statistical Modelling, 6*, 23-42.

O'Leary, C. and Eberts, R. W., (2008) The Wagner-Peyser Act and U.S. Employment Service: Seventy-Five Years of Matching Job Seekers and Employers. W.E. Upjohn Institute for Employment Research, http://bit.ly/2DpCEk3

Peterson, P. G., 1999. Gray Dawn: How the Coming Age Wave Will Transform America—and the World. Times Books, Random House: New York.

Porcellato L, Carmichael F, Hulme C, Ingham B, and Prashar A (2010) "Giving older workers a voice: Constraints on the employment of boomers in the North West of England." Work, Employment and Society 24(1): 85–103.

Rabe-Hesketh, S., Skrondal, A., and Pickles, A. (2005). Maximum likelihood estimation of limited and discrete dependent variable models with nested random effects. Journal of Econometrics, 128, 301-323.

Rabe-Hesketh, S., and Skrondal, A. (2012). Multilevel and longitudinal modeling using stata (3rd ed.). College Station, TX: Stata Press.

Resnick, B., Mills, M. E., and Nahm, E. (2004). "Web Use Can Improve Quality of Life for Elders," Maximizing Human Potential, Vol. 11, No. 3 (Winter 2004). pp. 4, 6.

Rix, S. (2006). "Foreword," in Charness, N. and Czaja, S. J. "Older Worker Training: What We Know and Don't Know," AARP working paper #2006-22.

Schmidt, F. L., and Hunter, J. (2004). "General Mental Ability in the World of Work: Occupational Attainment and Job Performance." Journal of Personality and Social Psychology, 86, 162-173.

Schneider, B. A., Daneman, M., Murphy, D. R., and Kwong See, S. (2000). "Listening to Discourse in Distracting Settings: The Effects of Aging." Psychology and Aging, 15, 110-125.

Schulz, R. and T. A. Salthouse. (1999). Adult development and aging: Myths and emerging realities (3rd Edition). Upper Saddle River, New Jersey: Prentice Hall.

StataCorp. (2015). Stata: Release 14 [statistical software]. College Station, TX: Author.

Sterns, H. L. and McDaniel, M. A. (1994). "Job Performance and the Older Worker." In Sara E. Rix (Editor), *Older Workers: How Do They Measure Up?* Washington, DC: AARP.

Thurow, L.C. (1975). *Generating Inequality*, New York: Basic Books.

U. S. Department of Labor, Bureau of Labor Statistics. (2005). *Tomorrow's Jobs.* Washington, D.C.: U.S. Department of Labor.

U.S. Department of Labor, Employment and Training Administration. (2007a). "PY 2005 WIASRD Data Book." Washington, DC: U.S. Department of Labor, Employment and Training Administration. http://bit.ly/2mMuBT4

U.S. Department of Labor, Employment and Training Administration. (2007b). "Senior Community Service Employment Program (SCSEP)." http://bit.ly/2DmhgvO

U.S. Department of Labor, Employment and Training Administration. (2008a). "PY 2006 WIASRD Data Book," Washington, DC: U.S., Department of Labor, Employment and Training Administration. http://bit.ly/2mM12RG

U.S. Department of Labor, Employment and Training Administration. (2008b). "PY 2007 WIASRD Data Book," Washington, DC: U.S. Department of Labor, Employment and Training Administration. http://bit.ly/2mOEXSn

U.S. Department of Labor, Employment and Training Administration. (2009). "The WIA National Summary of Annual Performance Data," Washington, DC: U.S. Department of Labor, Employment and Training Administration. http://bit.ly/2mNal3X

Weber P, Schaper M (2004) Understanding the grey entrepreneur. *Journal of Enterprising Culture* 12(2): 147–164.

Wooldridge, J. M. 2002. Econometric Analysis of Cross Section and Panel Data. Cambridge, MA: MIT Press.

Zellner, Arnold (1962). "An efficient method of estimating seemingly unrelated regression equations and tests for aggregation bias." *Journal of the American Statistical Association* 57: 348–368. doi:10.2307/2281644.

Zhang, T., (2008). *Elderly Entrepreneurship for an Aging U.S. Economy: It's Never Too Late*, London, New Jersey, Singapore: World Scientific.

Zhang, T. (2011), "Workforce Investment Act Training for Older Workers: Toward a Better Understanding of Older Worker Needs during the Economic Recovery," *US Department of Labor Employment and Training Administration Research & Publication,* No. ETAOP 2011-10.

Zhang, T. and Carr, D. (2014) "Does Working for Oneself, not Others, Improve Older Adults' Health? An Investigation on Health Impact of Self-Employment," *American Journal of Entrepreneurship*. Volume 7(1). Page 142–180. June 2014.

Zhang, T. (2014). "The Retired, the Resigned, or the Laid-off? Who Are the Urban Self-employed?" *Global Journal of Management and Business Research,* Volume 14 Issue 4. Page 7-14. May 2014.

Zhang, T. (2015) "What Drives Post-Retirement Age Self-Employment? An Investigation of Social, Policy and Individual Factors," in N. Sappleton and F. Lourenço (ed). *Entrepreneurship, Self-Employment and Retirement*, Hampshire, UK: Palgrave MacMillan. ISBN: 978-1-137-39837-6. April 2015.

Endnotes

[1] For WIA/WIOA adult programs, of those who are not employed at registration, it is defined as the number of adults who have entered employment by the end of the first three months after the exit quarter divided by the number of adults who exit during the quarter. For WIA/WIOA dislocated worker program, it is defined as the number of dislocated workers who have entered employment by the end of the first three months after the exit quarter divided by the number of dislocated workers who exit during the quarter.

[2] This survey is a combination of a team effort between the Society for Human Resource Management (SHRM), the National Older Worker Career Center (NOWCC), and the Committee for Economic Development (CED). A sample of HR professionals was randomly selected from SHRM's membership database, which consists of more than 170,000 members. In November 2002, 2,500 randomly selected SHRM members received an e-mail invitation containing a link that directed them to the online survey. Of these, 2,143 e-mails were successfully delivered to respondents, and 428 HR professionals responded, yielding a response rate of 20%.

[3] The aging workforce is projected to result in potential labor force shortages, Social Security fund bankruptcy and other related fiscal pressure, under the current economic and technological conditions. Older individuals' participation in the labor force can be a possible solution to those above socioeconomic problems (Zhang 2008). The Great Recession and decimated retirement assets in it are also pushing seniors to remain or return to the labor force. Retaining seniors in the labor force becomes a necessity for both older individuals and the economy.

[4] The Program Years span respectively from April 2013 to March 2014, April 2014 to March 2015, and April 2015 to March 2016.

[5] In this table calculated by the (ETA (2017a), older workers are defined as workers age 55 and above. However, the study described in this paper defines older workers as age 50 and above, to include more older workers.

[6] Adults include all adults who are age 18 and above. The priority for intensive and training services are given to low-income individuals and individuals receiving government assistance and with limited resources (ETA 2017d). Dislocated workers instead refer to individuals who have been or to be terminated or laid off (particularly for business closures or structural unemployment reasons), who are eligible for or have exhausted unemployment insurance, who are unlikely to return to a previous industry or occupation, or who are displaced homemakers who are no longer supported by other family members (ETA 2017d). For them, job placement assistance and skill assessments and counseling could be helpful.

[7] Using entered employment rate to measure performance may cause some concerns. First, the measure is based on program exit. Focusing exclusively on exiters could introduce selection bias. One suggestion could be to focus on program entry as an alternative measure. However, using entry as a measure offers no clue on when exit occurs and cannot measure participants' post-program employment outcomes. However, with an appropriate measure of program participation length as a variable, future sensitive study comparing using entry and exit as a measure for performance could be interesting. Second, the outcome measure, EER, is measured only one-quarter after program exit. Many workers might not find jobs until several quarters after program entry/exit and the program effect might not show up until then. However, at the moment, the one-quarter lag for the performance measure, EER, is still a commonly used measure for WIOA services, reflected in the WIASRD data. Again, future investigation on the optimal length of observing lags could be a possible extension of the research.

[8] WIB refers to Workforce Investment Board.

[9] The author initially considered using Seemingly Unrelated Regression model (SUR) developed by Arnold Zellner (1962). SUR is a technique for analyzing a system of multiple equations with cross-equation parameter restrictions and correlated error terms. In this study, the model contains two independent equations. Each of those two equations used a different and unrelated dataset, due to the very different attributes between dislocated worker participants and adult participants in the WIA programs. For example, participants in the adult program have extremely low wages and very little prior attachment to the labor market, while participants in the dislocated worker program have had strong attachment to the labor market and relatively high wages. These two participants groups do not have necessary relationships. In this case, the SUR assumption that error structures of the two models are similar does not necessarily hold. There is no need to estimate SUR; instead, the models for adults and dislocated workers are estimated separately.

[0] Please note that total numbers of older dislocated workers and older adults are not the same as the number of observations shown later in the regression models because not all worker information is reported for all variables used in the regression models.

[1] Local Veterans Employment Representative.

[2] Disabled Veterans Outreach Program.

[3] Please note that the service participation length does not mean a participant's total participation in a specific WIA program. Instead, it is calculated as the time span from the participation of the first service to the exit from the last service. This would therefore result in some significant program participation length because of returning participants for the same or different services over the years. Based on preliminary analysis on training length, it was associated with the nature of service/training and other personal factors like the participants' education attainment.

[4] Please also note that many older workers reported in the WIASRD data did not offer occupational information. Therefore, when estimating our models incorporating occupational categories, many observations were dropped. However, this situation happened not just to this variable; a few other variables faced the challenge of

missing values as well. This occupational training variable is one of the key observing variables and it is an important control variable due to different nature of occupational trainings. Even with dropped observation, it is still necessary to be included.

III. LERA Best Posters

Institutional Change in the Apprenticeship Systems of England and Germany

MAISIE AUFDERHORST-ROBERTS[1]
Leeds University Business School

The English and German apprenticeship systems face change, yet the underlying causes and future implications of recent developments remain unclear. Conventional wisdom views the English and German systems as polar opposites. The "dual" and corporatist German system provides highly structured training whilst England's employer-led system prioritizes the market-led flexibility of the economy. Despite this, recent developments highlight convergence, with higher investment in apprenticeships forecast in England compared to an increasing provision of higher education in Germany. However, empirical evidence indicates that the systems are not converging. Instead, adopting an institutional perspective is essential to understanding current and future policy developments.

"It All Revolved Around Numbers": Greater Commodification of Work and Culture with Outsourcing

Jacqueline M. Zalewski[2]
West Chester University

Private sector unions are in decline. Their future is dependent on engaging workers to the point of active participation. Previous research has demonstrated that such participation is preceded by the development of some level of member loyalty. Our study examines key antecedents to union loyalty and identifies a key mechanism through which these factors work. Specifically, we find pro-union attitudes and union instrumentality to be significant predictors of union member loyalty. Additionally, procedural justice perceptions mediate the relationship between antecedents and loyalty. Thus, our findings reflect the view that union socialization may be important in developing member loyalty.

[1] Leeds University Business School, Maurice Keyworth Building, The University of Leeds, Leeds, LS2 9JT, United Kingdom
[2] Department of Anthropology and Sociology, West Chester University, 700 South High Street, West Chester, PA 19383

IV. LERA Competitive Papers II

Unpacking Job Satisfaction and Union Participation: The Role of Fit

JUSTIN P. WIEGAND
ROBERT BRUNO
University of Illinois at Urbana-Champaign

Job satisfaction's role as an antecedent to union participation has often been proposed—generally as a negative relationship—but empirical support is lacking. To clarify boundary conditions of this relationship, we turn to the exit-voice tradeoff, the attraction-selection-attrition framework, and the frustration-aggression hypothesis. We suggest a negative job satisfaction-union participation relationship exists only among workers lacking fit with their colleagues ("person-workgroup fit"). We employed a distance-based measure of person-workgroup fit to analyze data from 777 workers across three unions (90% public-sector) located in a large Midwestern city. Results indicate fit's moderating role—high-fit workers participate in union activities irrespective of their job satisfaction, but workers with low fit participate more when dissatisfied with their jobs. Our findings inform theory on antecedents of union participation and the strategic choices unions face in organizing and reinvigorating lay activism.

> *It's a long struggle to change an engrained practice where people don't participate, and don't expect to participate, and don't expect their participation to make a difference.*
>
> —Union officer quoted in Hickey (2005, p. 290)

This article empirically examines the relationship between union members' job satisfaction and union participation conditional on one's similarity, or fit, with their workgroup. Although job satisfaction's role as an antecedent to union participation has been well studied, important questions remain unanswered. Lay participation is essential to unions' viability and growth, yet as noted in the opening quote, increasing participation is hard fought. Understanding all antecedents of union participation is thus important to both unions and labor scholars. Job satisfaction's specific relationship to union participation has generally been thought to be negative, but this proposition lacks empirical support. Although Bamberger, Kluger, and Suchard (1999) established a meta-analytic corrected correlation (ϱ) of union participation and job satisfaction of $-.16$ ($p < .05$, k = 17), correlations became more disparate as the literature expanded. Monnot, Wagner, and Beehr's (2011) updated meta-analysis (k = 31) ultimately found a nonsignificant relationship between the two variables ($\varrho = -.04$).

Although Monnot and colleagues' finding suggests moving away from viewing job satisfaction as a union participation antecedent, simple correlational models may not accurately capture the job satisfaction-union participation relationship. Understanding if, when, and how a negative job satisfaction-union participation relationship exists would inform union organizing efforts and models of participation. Indeed, rank-and-file participation and voluntary activism serve as the "very fabric of unions" (Gordon, Beauvais, and Ladd, 1984, p. 480). Knowing workers most likely to require encouragement to participate in union activities

Corresponding author's address: Justin Wiegand, School of Labor and Employment Relations, University of Illinois at Urbana-Champaign, 504 E. Armory Avenue, Champaign, IL 61820 helps to focus union outreach and theoretically-derived moderators of the job satisfaction-union participation relationship are worth consideration. Naturally, if certain members participate more when they are less satisfied, unions should consider how to engage these workers and seek to understand why dissatisfied workers are generally their greatest participants. To this end, we believe accounting for "misfits," those who do not share the interests and values of their workgroup, stands to determine whether a negative relationship exists between job satisfaction and union participation.

We suggest that workers most similar to their colleagues experience "natural" solidarity and stay involved in their union regardless of job satisfaction. Consider one union operating engineer who told a study author that working with "like-minded union brothers" provided a sense of belonging and pride motivating his union involvement (personal communication, September 10, 2016). Misfits who do not share the interests of their workgroup might still participate in their union, but for additional reasons. Specifically, job satisfaction stands to meaningfully relate to union participation for these individuals. Consider a union teacher who had not immediately connected with her union brothers and sisters but was driven to participate to "protect [her] profession" from being treated as "cogs in a wheel" (personal communication, April 4, 2013). Thus, although high-fit union members may participate regardless of their satisfaction, misfits are different. Those not satisfied at work are likely to either leave or participate in their union as an alternative to exit. In other words, dissatisfied misfits persisting with an employer are likely to participate well with their union. Satisfied misfits represent a unique subgroup. Neither fit nor dissatisfaction motivate misfits' participation leading us to predict that they less likely to participate in union activities as a group. In the remainder of this article, we develop these ideas in view of the exit-voice tradeoff, the attraction-selection-attrition framework, and the frustration-aggression hypothesis.

We used primary data from members of three unions (90% public-sector) operating in a large, Midwestern city, to evaluate how members' fit with their workgroup determines when a negative relationship exists between job satisfaction and union participation. This moderation analysis was performed using a distance-based measure of interest fit to compare both individual union members' interests and the general interests characterizing their workgroups. We found a negative relationship between job satisfaction and union participation when fit was low, but no relationship when fit was high. Instead, participation remained relatively high for high-fit individuals regardless of their level of job satisfaction. Our accounting of fit suggests a critical boundary condition for the job satisfaction-union participation relationship and offers an explanation for prior inconclusive findings. Our findings also inform the strategic choices unions face in reinvigorating participation and lay activism.

Theory, Literature, and Hypotheses

Antecedents of union participation have been studied extensively by labor scholars culminating in two meta-analyses on the subject (Bamberger et al., 1999; Monnot et al., 2011) that found positive relationships with pro-union attitudes, union instrumentality perceptions, union commitment, and organizational commitment. Other antecedents have been proposed and substantiated thereafter (e.g., perceived behavioral control; Fiorito, Padavic, and Russell, 2014). Job satisfaction's role as an antecedent to union participation has often been proposed and evaluated as well. Yet, as reviewed previously, consistent empirical support is lacking.

Theoretical backing for a negative job satisfaction-union participation relationship follows from several theoretical developments including the frustration-aggression hypothesis, exit-voice trade off and the attraction-selection-attrition framework. The frustration-aggression hypothesis (Klandermans, 1986; Wheeler, 1985) considers union participation to be "a reaction to frustration, dissatisfaction, or alienation in the work situation" (Klandermans, 1986, p. 190). Indeed, those lacking job mobility, perceiving their wages as low, and generally feeling inequality with their employer have been demonstrated to be more likely to participate in their unions (Huszczo, 1983; Kolchin and Hyclak, 1984). Yet, beyond single studies, lacking meta-analytic evidence and wide job satisfaction-union participation credibility intervals (−0.28 to 0.19; Monnot et al., 2011) suggest a more nuanced model is needed to explain job satisfaction's relationship with union participation.

The exit-voice tradeoff extends the proposed negative relationship between job satisfaction and union participation (i.e., voice) to include exit. The exit-voice tradeoff suggests workers respond to discrepancies between desired and actual environmental circumstances by either exiting the undesirable situations or by expressing dissatisfaction through "voice." This offered an explanation for lower mean job satisfaction among union workers compared to non-union workers (Borjas, 1979; Freeman, 1980). In line with the present study, the exit-voice tradeoff has been extended to union-only samples by evaluating the relationship between an individuals' job satisfaction and the extent to which they participate in their union (e.g., Iverson and Currivan, 2003).

Schneider's (1987) attraction-selection-attrition (ASA) framework provides additional insight on when workers persist in or exit firms. The ASA framework suggests people are attracted to and select themselves into and out of organizations. People who select into and persist in an organization "make the place" by normalizing organizational culture and function. As a whole, the ASA framework suggests individuals with interests and values similar to those held by members of an organization are more likely to select into and persist in that given organization. The ASA framework can be extended to workgroups as well, such that individuals are more likely to select into and persist in workgroups that match their own interests and values.

Taken together, we suggest person-workgroup (PW) fit determines whether job satisfaction and union participation are negatively related. PW fit refers to the similarity of a worker's interests to the interests generally held by his or her colleagues, specifically those working in similar occupations within a given organization. This follows Holland's (1966) idea that environments reflect the people in them. In his words, "the *dominant* features of an environment are dependent upon the *typical* characteristics of its members" (p. 53, emphasis in original). Holland (1997) eventually operationalized occupational environments as the distribution of personality types in a given occupation, an idea we extend to workgroups.

Theories of person-environment (PE) fit, including the ASA framework and Holland's conceptualization of interest fit, describe individuals as preferring and seeking out environments compatible with personal characteristics (Kristof-Brown, Zimmerman, and Johnson, 2005; Schneider, 1987). Individuals in compatible environments usually exhibit desired personal and organizational outcomes, including reduced stress, higher job satisfaction and performance, and lower turnover (Kristof-Brown et al., 2005; Nye, Su, Rounds, and Drasgow, 2012).

PW fit provides a natural application for considering PE fit in unionized settings. A worker's unionized colleagues provide an immediate, day-to-day connection to his or her union. Thus, when workers share the interests and values of their workgroup (i.e., experience high PW fit) they have reason to involve themselves in their union apart from their level of job satisfaction. Union participation provides connection and camaraderie; solidarity can form organically. However, as a whole, PW misfits do not enjoy the same connection to their colleagues as those with high fit. Camaraderie does not develop as easily and solidarity does not follow so naturally. Thus, to the degree misfits find satisfaction in their work, they are less likely to participate in their union absent intentional union efforts to engage and include them.

To fully understand our propositions, consider the experiences of workers high and low on both PW fit and job satisfaction. Individuals with high fit, but low satisfaction are relatively likely to participate in their union. High fit suggests they have similar interests as their colleagues and are likely to find their work engaging and interesting, but low satisfaction indicates there may be circumstances and stressors related to their work that they want to see change. Union participation provides a means to promote change to move their unsatisfying work toward a place of satisfaction. Participating in their union affords them voice to achieve this end while connecting with the colleagues they fit so well with.

Individuals with high fit *and* high job satisfaction are also relatively likely to participate in their union. As before, high fit again suggests they have similar interests to their colleagues and are likely to find their work engaging and interesting. Their high job satisfaction indicates agreeable work circumstances. Participation with their fellow union members follows high fit with their workgroup and enjoyment of the work itself. Following the ASA framework, those with similar interests as their workgroup most naturally select into and persist in the workgroup—solidarity forms naturally and is stoked by union participation.

Taken together, union participation for high-fit individuals should remain relatively high regardless of union members' job satisfaction. Similarity to their workgroup promotes solidarity and union involvement at all levels of job satisfaction. Individuals can participate in their union to promote change, or simply to connect with like-minded colleagues, a byproduct of sharing the dominant vocational interests of their workgroup. Thus:

Hypothesis 1: There is little relationship between job satisfaction and union participation for high PW fit union members

Following the exit-voice tradeoff and the ASA framework, low-fit workers with low job satisfaction are likely to leave their employer unless they can change their work or working conditions. Participating in their union provides them the voice alternative to exit the rather grim circumstance they find themselves in—poor fit and dissatisfying work. The reality of both poor fit and low satisfaction suggests these individuals who persist with their employer will be especially motivated to participate in their union to promote change, regardless of their misfit status among colleagues. The exit-voice tradeoff describes low-fit, low-job satisfaction individuals well. Although neither fit nor satisfaction compels them to stay in their job, voice (i.e., union participation) can empower them to stay.

On the other hand, individuals with low fit, but high job satisfaction are less likely to participate in their union than their colleagues. Although generally satisfied with their work, their low fit suggests a relative disconnect from colleagues holding the dominant vocational interests of their workgroup. The reality that their job satisfaction remains high suggests these workers see little need for change at work. Since camaraderie and solidarity are also not expected to occur as naturally, union participation should be relatively low for satisfied misfits.

When both low-fit scenarios are taken together, it becomes clear that a negative job satisfaction-union participation relationship should be observed for low-fit workers. The exit-voice tradeoff is descriptive of low-fit individuals who are not satisfied with their work. These workers are the most likely to either leave their firm or participate in their union. Thus, low-fit workers with low job satisfaction who remain with their employers should participate at higher than average levels with their union. On the other hand, union participation from low-fit workers with high job satisfaction should be low—solidarity comes less naturally for misfits and they have less motivation to seek workplace change via union participation since they are already satisfied with their work. Thus:

Hypothesis 2: The relationship between job satisfaction and union participation is negative for low PW fit union members

Figure 1 depicts the path analysis underlying our hypotheses. To summarize, our conceptualization of the job satisfaction-union participation relationship follows relevant developments on the antecedents of union participation and are grounded in Schneider's (1987) attraction-selection-attrition (ASA) framework and the within-union application of the exit-voice tradeoff (e.g., Iverson and Currivan, 2003). We suggest that PW fit moderates the job satisfaction-union participation relationship. Those who fit with their workgroup have reason to participate in their union regardless of their level of job satisfaction, but low-fit union members who are not satisfied are less likely to participate. To the degree such cases exist, they represent opportunities for unions. At a time when union membership and lay activism are waning and hard fought, unions should double-down in their efforts to identify and engage disenfranchised workers.

FIGURE 1
Hypothesized Path Model Distinguishing Exit-Voice Tradeoff in Union Samples

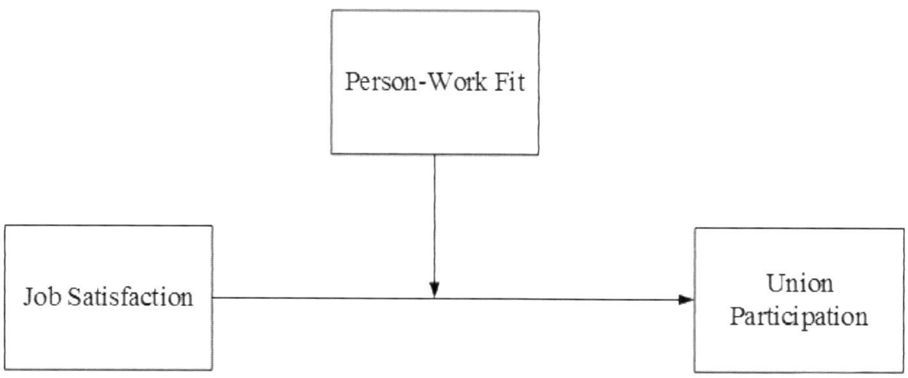

As an example of fit's moderating role of the job satisfaction-union participation relationship, consider teachers (the largest workgroup in our sample). Misfit K – 12 teachers may include teachers hired through the Teach for America (TFA) program. TFA teachers are drawn from non-educational academic backgrounds and have not undergone teaching preparatory programs ("Is TFA for You?," 2016). Most of the TFA recruits have non-educational career interests and are pursuing degrees in applied fields. Unlike teachers from undergraduate educational programs, TFA "corps" members enter short two-year teaching commitments and most then move on to their preferred vocational path. Leadership is a strongly sought after characteristic of TFA recruits and most descriptive of individuals high on Holland's (1997) enterprising interest, an interest not typically highest for teachers. If such misfits are ultimately satisfied with their work as teachers, their vocational disconnect from other union members might preclude them from participating fully in their union.

Data and Analysis

In 2015, we invited leaders of seven unions located in a large Midwestern city to ask members of their respective unions to participate in our study. Leaders from five unions representing several industries and both the private and public sectors, agreed to participate. These unions' members were emailed directly by their respective leadership with an invitation to complete our study's linked online survey. Two of the five unions only yielded a single response and were excluded from the sample. The three unions making up our sample include one public- and two private-sector unions providing 1,232 respondents.

All respondents were located in a Midwestern, free collective bargaining (i.e., non-right-to-work) state. The clear majority of individuals in the final sample (90%) were members of the public-sector union. This union represents employees in a large, metropolitan school district. Noted differences exist between private and public unions, including the inability of many public-sector unions to strike. In the present case however, the public-sector union was in a strike-permissive state. Indeed, the studied union struck in a previous bargaining period and passed a strike-vote in the most recent bargaining period, narrowly avoiding a strike occurrence. All considered, our results hold relevance for both public-sector unions in strike-permissive states and private-sector union members. Furthermore, our results were robust to public-/private-sector controls.

Measures

Vocational Interests

Holland's (1959, 1997) model of interests provides an operational basis for the person and environment components of PW fit. Holland described six interests, realistic, investigative, artistic, social, enterprising, and conventional, which are often referred to by their acronym, RIASEC. Table 1 describes work preferences and occupations characteristic of each interest. The interrelatedness of RIASEC interests is often depicted using a hexagon (see Figure 2; e.g., Holland, 1997; Tracey and Rounds, 1993): adjacent interests (e.g., realistic and conventional) are more related than alternate interests (e.g., realistic and enterprising), which are more related than opposite interests (e.g., realistic and social).

Person interests were measured using the Department of Labor's public-domain Occupational Information Network (O*NET) Interest Profiler Short Form (Rounds, Su, Lewis, and Rivkin, 2010). The form is a 60-item measure with ten items corresponding to each RIASEC interest. Respondents were asked to decide the degree to which they would like or dislike doing a type of work regardless of whether they had education or training to do the work, how it was related to their current job, or how much money they would make doing the work. Responses ranged from −3 ("Dislike very much") to 3 ("Like very much") with a neutral midpoint (0 = "Neither like nor dislike"). "Lay brick or tile" represents a realistic item and "perform rehabilitation therapy" a social item. Scale scores were formed by taking the mean of all scale items. Reliabilities (Cronbach's α) ranged from .82 (social) to .92 (investigative) for RIASEC scale scores.

We evaluated the environmental interests characteristic of the following occupation groups: teachers, clinicians, and paraprofessionals within a metropolitan teachers union, operators from a local operating engineers union, and wiremen from an electrical workers local. The teachers' group included kindergarten to twelfth grade teachers, including special education faculty. Clinicians included counselors, social workers, physical and occupational therapists, and speech pathologists. Technicians, assistants, clerks, and secretarial staff made up the paraprofessionals group. Heavy equipment operators and mechanics made up the group of operators and apprentice and journey wiremen and electricians comprise the group of wiremen.

TABLE 1
Holland's (1997) RIASEC Interests and Example Occupations

Interest	Individuals with this interest seek and prefer…	Occupations
Realistic	Activities involving the "manipulation of objects, tools, machines, and animals" (p. 43).	Bricklayer, electrician, plumber
Investigative	The "creative investigation of physical, biological, or cultural phenomena" (p. 44).	Chemist, psychologist, biologist
Artistic	"Ambiguous, free, unsystematized activities and competencies to create art forms or products" (p. 45).	Graphic designer, actor, craft artist
Social	"Inform[ing], train[ing], develop[ing], cur[ing], or enlighten[ing]" others (p. 46).	Nurse, teacher, clergy
Enterprising	"Attain[ing] organizational or self-interest goals" (p. 46).	Retail salesperson, Lawyer, CEO
Conventional	The "systematic manipulation of data such as keeping records, filing [and reproducing] materials, organizing written and numerical data according to a prescribed plan, [and] operating business and data processing equipment" (p. 47).	Accountant, librarian, editor

FIGURE 2
Holland's RIASEC Model of Interests with Prediger's People/Things and Data/Ideas Dimensions

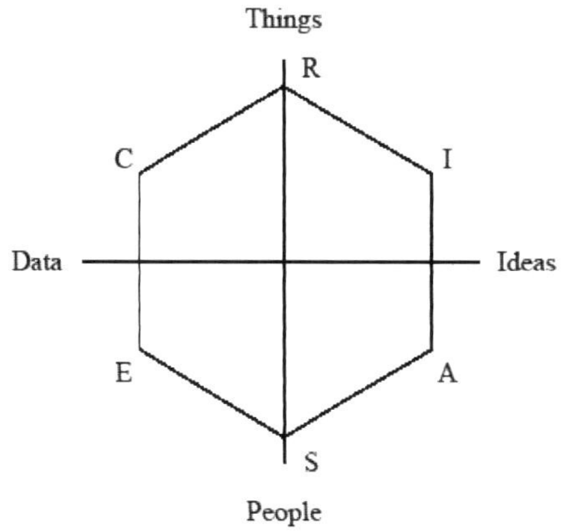

Occupational interest profiles were formed from the mean RIASEC scores of all respondents within a given occupational group. As in previous studies of interest fit (e.g., Su, 2012), occupational interest profiles for each individual were formed from the mean of all RIASEC scores except the individual's own so as not to inflate the subsequent evaluation of fit. This leave-one-out aggregation technique provides slightly different environmental interest profiles for each individual within a given occupational group. By evaluating person and environment separately, and with an aggregated environment, we are able to calculate an indirect, objective measure of fit (Kristof-Brown and Guay, 2011) reducing common method bias in our results (Podsakoff, MacKenzie, Lee, and Podsakoff, 2003).

Person-Workgroup Fit

Since Holland (1963) first recommended comparing first letters of person and environment RIASEC profiles for agreement, numerous indices have been introduced to measure RIASEC fit. Camp and Chartrand (1992) and Brown and Gore (1994) offer reviews of many RIASEC-specific fit indices. Unfortunately, most of the RIASEC fit indices developed fail to consider entire RIASEC profiles. Zener and Schnuelle (1976) eventually expanded fit indices to consider the first three letters of person and environment RIASEC interest scores, but subsequent indices generally failed to move past this three letter comparison threshold. Profile correlations allow for a comparison of all six scores, but fail to directly account for scale-level differences between person and environment. In other words, so long as person and environment scores correlate with each other, fit scores are high, regardless of the similarity or difference of relative strengths of each component measure. Calculating the Euclidean distance between person and environment scores based on all RIASEC interests accounts for scale-level differences by comparing the distance between all person and environment scores. We operationalized fit as the Euclidean distance between person and environment scores due to this superior property and the parsimony of a single fit metric.

Euclidean distance calculations of person and workgroup RIASEC scores followed Tracey, Allen, and Robbins' (2012) calculations. First, complete person and environment RIASEC profiles were transformed to points in Prediger's (1982) two-dimensional People-Things (PT) and Data-Ideas (DI) space. This space represents a parsimonious conception of Holland's hexagon by transforming the six RIASEC interests into a single point on these two dimensions. The two dimensions can be directly overlaid on the RIASEC hexagon (note their axes in Figure 2).

The PT dimension was calculated as $2*R + I - A - 2*S - E + C$ such that positive values represent Things interest and negative values represent People interest (letters in the equation represent the first letter of each interest). The DI dimension was calculated as $1.73*E + 1.73*C - 1.73*I - 1.73*A$ such that positive values represent Data interest and negative values represent Ideas interest. Euclidean distance was then calculated as:

$$distance = \sqrt{(PT_{person} - PT_{environment})^2 + (DI_{person} - DI_{environment})^2}$$

Very high raw distance values represent low fit (i.e., the person and workgroup scores are far apart) and very low raw values (i.e., approaching zero) represent high fit (i.e., the person and workgroup scores are close together). To aid in interpretability, we multiplied the raw Euclidean distance calculations by −1 and mean centered the results. Thus, negative values represent below average fit in the study sample and positive values represent above average fit.

Job Satisfaction

Although job satisfaction can be evaluated at a facet-level (Judge, Hulin, and Dalal, 2012; Locke, 1969; Spector, 1997), studies evaluating overall job satisfaction generally use either Brayfield and Rothe's (1951) five-item scale or the Job in General Scale (Ironson, Smith, Brannick, Gibson, and Paul, 1989) and its abridged version (Russell et al., 2004). In all cases however, these scales were not created with union environments expressly in view. In the present study, we desired a union-specific measure of overall job satisfaction—a measure that would account for overall job satisfaction including the unique concerns of unionized workers.

We developed job satisfaction items that assess the concerns of union workers in their vernacular. These items were reviewed and refined with feedback from union members and labor studies faculty before being administered. Respondents were asked to evaluate their agreement with items (sample item: "Since starting my job, I have less time to do the job I was hired to do") using a seven-point scale (−3 = "Strongly Disagree," 3 = "Strongly Agree") with a neutral midpoint (0 = "Neither Agree nor Disagree"). Scale items are provided in the appendix.

As a comparison point, Brayfield and Rothe's (1951) scale was also administered in the present study. Our new scale correlated .60 with Brayfield and Rothe's measure (using the same seven-point scaling), suggesting we are measuring a similar, but not identical, job satisfaction construct. Scale reliability (Cronbach's α) for our union-specific measure was .88.

Union Participation

Scholars differ on the dimensionality of union participation. Some use a single factor (e.g., Anderson, 1979) and others suggest multidimensional models (e.g., McShane, 1986; Monnot et al., 2011; Parks, Gallagher, and Fullagar, 1995). A distinction is most often made between nonmilitant participation, which involves activities that do not require direct interference with work (e.g., helping with union organizing activities), and militant participation, which involves high-intensity activities interfering with one's work (e.g., striking). As both types of participation are relevant to the current study, we sought a parsimonious measure with items addressing both militant and nonmilitant participation.

Existing scales of both militant (e.g., Martin, 1986) and nonmilitant (e.g., McShane, 1986) union participation use a mix of Likert, dichotomous, and open-ended numeric items (e.g., "number of union meetings attended"). The mix of scaling within and between instruments makes scale-level interpretation difficult. Instead of combining such items, we began with the existing scales and again drew on the experience of labor studies faculty and union members to create a Likert scale assessing overall union participation. Nine items comprised our scale of overall union participation; they are provided in the appendix. Two items (marked with asterisks) were adapted directly from Martin's (1986) Militancy scale.

Our scale covered general participation in union activities and service along with militant actions, both legal and those involving acts of civil disobedience or illegal work stoppages. Respondents were asked to evaluate their agreement with four items (e.g., "I would never engage in violence during a strike"; reverse-coded) using a seven-point scale (−3 = "Strongly Disagree", 3 = "Strongly Agree") with a neutral midpoint (0 = "Neither Agree nor Disagree") and their participation relative other employees in their union with five items (e.g., "I help with union organizing efforts"), also using a seven-point scale (−3 = "Extremely Below Average", 3 = "Extremely Above Average") with a neutral midpoint (0 = "An Average Amount"). Scale reliability (Cronbach's α) was .82.

Controls

We controlled for differences individuals' employment sector, status as a racial minority, and level of education. The public–private sector control provides guidance on whether our findings generalize beyond the public-sector union comprising much of our sample. The racial majority control (1 = racial majority) informs whether fit and misfit with the racial in-group should be considered in our fit conceptualization. Finally, the ordinal accounting of education level (1 = "some high school" to 7 = "doctorate") seeks to account for education and education-related status differences in union participation.

Analysis

We estimated the effect of job satisfaction on union participation conditional on fit using SAS 9.4 (*SAS*, 2012). The full equation can be written as:

$$Participation = b_0 + b_1(JS) + b_2(Fit) + b_3(JS \times Fit) + b_4(RM) + b_5(Edu) + b_6(\mathrm{Sec}) + e$$

where "JS" represents job satisfaction, "RM" is a racial majority group dummy, "Edu" represents education level, "Sec" is a public-sector employment dummy, and "e" represents the error term. Simple slopes of union participation at plus and minus one standard deviation of fit and job satisfaction were calculated and plotted along with their 95% confidence intervals following Aiken and West (1991). A significant interaction between job satisfaction and fit (b_3) and a negative simple slope estimate of job satisfaction and union participation only when fit is low would support our hypotheses.

Empirical Findings

Table 2 provides descriptive statistics and correlations of our study variables. Fit is mean centered, and job satisfaction and union participation are both centered on their neutral scale midpoint to aid in interpretability. Respondents who correctly answered at least four of five randomly placed quality control items (e.g., "Please select 'Strongly Agree.'") in our survey comprised our final sample of 777 union members (70% female and 30% racial minorities). Ninety percent held a bachelor's degree or higher and 71% held a master's degree or higher. Respondents' mean age and organizational tenure was 43.46 (SD = 11.11) and 11.93 (SD = 8.70) respectively. Mean job tenure and current union tenure was 9.63 (SD = 8.60) and 13.22 (SD = 9.35) respectively. Consistent with the job satisfaction-fit literature, individuals' job satisfaction positively related with their PW fit ($r = .09$, $p < .05$). Multidimensional scaling (Borg and Groenen, 2005) performed in SAS 9.4 (*SAS*, 2012) verified the structure of RIASEC interests for individuals. As shown in Figure 3, the ordering of the RIASEC scales followed the circular ordering of Holland's hexagon.

Overall, the characteristic interests of workgroups followed expectations. RIASEC means and People-Things/Data-Ideas calculations for each of the five workgroups are shown in Table 3. Following O*NET interest profiles of occupations underlying our workgroups (Peterson, Mumford, Borman, Jeanneret, and Fleishman, 1999), we expected teachers, clinicians, and paraprofessionals to all be high on social interest. Indeed, the mean score for social interest was highest for these workgroups and exhibited the least variability (S.D.) across individuals in those workgroups.

TABLE 2
Descriptive Statistics and Correlation Matrix of Study Variables

Variable	Mean	S.D.	Range	1	2	3	4
1. Job Satisfaction	−0.80	1.45	−3 – 3	.88			
2. Realistic	−0.61	1.37	−3 – 3	.07*	.90		
3. Investigative	0.32	1.48	−3 – 3	−.02	.57*	.92	
4. Artistic	0.84	1.29	−3 – 3	−.05	.21*	.35*	.88
5. Social	1.36	0.99	−3 – 3	−.05	.08*	.25*	.38*
6. Enterprising	−0.14	1.22	−3 – 3	.00	.28*	.30*	.29*
7. Conventional	−0.44	1.34	−3 – 3	.02	.41*	.30*	.10*
8. PW Fit	−0.01	2.82	−11.16 – 5.23	.09*	.04	.07	.04
9. Union Participation	−0.27	1.06	−3 – 3	−.09*	.22*	.16*	.08*
10. Racial majority[a]	0.70	.46	0 – 1	−.03	.07	.01	−.08*
11. Level of education	5.59	.84	1 – 7	−.30*	−.24*	−.03	.12*
12. Public sector[a]	0.90	.31	0 – 1	−.23*	−.31*	−.07	.15*

Note. N = 759 – 777. PW = Person-Workgroup. Reliabilities (Cronbach's α) are provided on the diagonal when applicable.

[a]1 = Yes.

*$p < .05$.

FIGURE 3
Two-Dimensional Scaling of Person RIASEC Interests
(the first letter of each point corresponds to the first letter of a respective RIASEC interest)

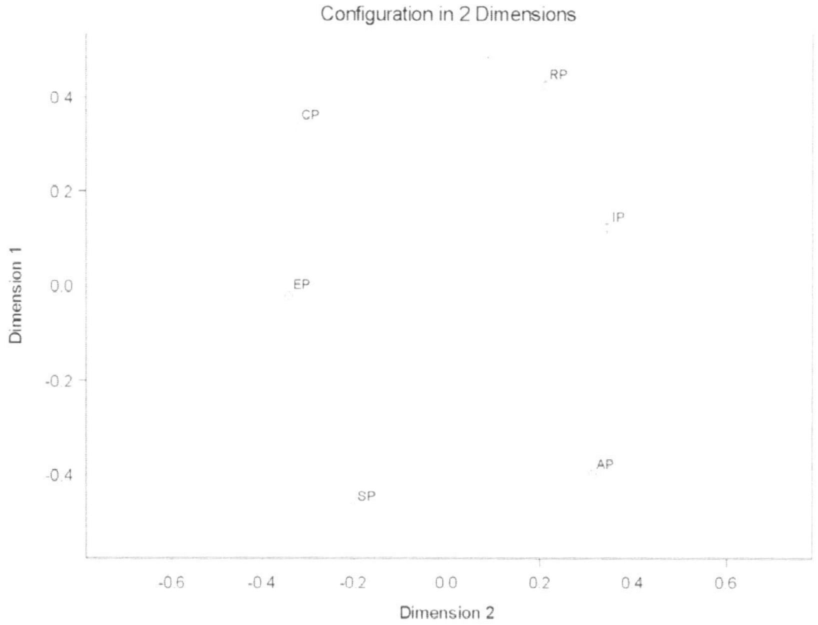

TABLE 3
Workgroup Interest Profiles

	Teachers	Clinicians	Paraprofessionals	Operators	Wiremen
Realistic	−0.73	−1.27	−0.44	0.31	0.81
Investigative	0.29	0.07	0.60	0.17	0.85
Artistic	0.90	0.69	1.51	0.23	0.27
Social	1.43	1.76	1.64	0.47	0.50
Enterprising	−0.18	−0.06	0.73	−0.29	−0.08
Conventional	−0.47	−0.63	0.89	−0.89	−0.28
People-Things[a]	−5.22	−7.25	−4.91	−0.98	1.00
Data-Ideas[b]	−3.18	−2.51	−0.85	−2.73	−2.56
N	621	51	24	27	54

Note. People-Things and Data-Ideas calculations are provided in the text.

[a]Positive values represent Things.

[b]Positive values represent Data.

Wiremen and operators were both expected to be high on realistic interest and were. Compared to the teacher's union groups, whose average realistic scores represented a "slight dislike," wiremen and operators both exhibited positive mean realistic scores representing a degree of liking. The mean score for investigative interest was slightly higher than realistic interest for wiremen, but the practical significance of this difference was trivial (.04 scale points). For operators, the mean score for social interest was slightly higher than realistic, but again, the practical significance of this difference was very small (.16 scale points). Figure 4 plots each of the workgroups in Prediger's (1982) two-dimensional People-Things/Data-Ideas space. Workgroups' People-Things/Data-Ideas scores followed expectation as teachers, clinicians, and paraprofessionals all exhibited stronger People scores than operators and wiremen.

FIGURE 4
Workgroup Interest Profiles on People-Things, Data-Ideas Dimensions

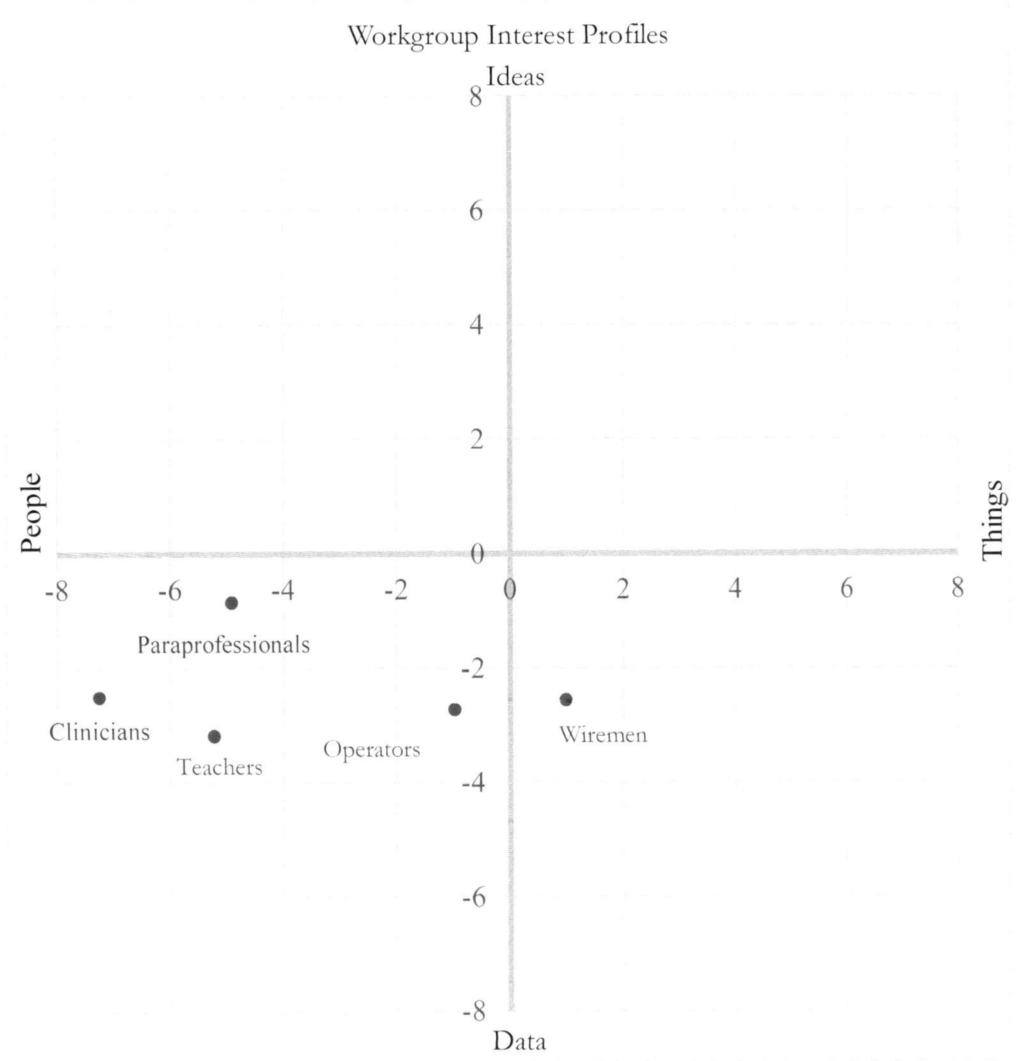

Moderation results supported our hypotheses. Table 4 provides results of the regression analysis with and without controls. In both cases, the job satisfaction-PW fit interaction was significant as well as the negative simple slope of job satisfaction and union participation when fit was low (−1 S.D.). The relationship between job satisfaction and union participation when fit was high (1 S.D.) was not significant. Table 5 provides the effects of these simple slope results and Figure 5 displays a plot of the slopes and 95% confidence intervals for job satisfaction and fit without controls. Low- and high-fit regions are significantly different for all positive job satisfaction scores. Notably, our job satisfaction scale explained 71% more variation of the data in our hypothesis tests (R^2) than the Brayfield and Rothe measure. The increase in explained variance suggests an improvement in the criterion-related validity of job satisfaction in union samples.

TABLE 4
Determinants of Union Participation

	Without Controls		With Controls		
	Model 1	Model 2	Model 3	Model 4	Model 5
Intercept	−0.33* (.04)	−0.34* (.04)	−0.14 (.29)	0.07 (.29)	0.02 (.29)
Job Satisfaction	−0.07* (.03)	−0.07* (.03)		−0.11* (.03)	−0.11* (.03)
PW Fit	0.02 (.01)	0.05* (.02)		0.02 (.01)	0.03* (.02)
JS x PW Fit		0.02* (.01)			0.02* (.01)
Racial majority (1 = yes)			0.03 (.08)	0.02 (.08)	0.02 (.08)
Level of education			0.12 (.07)	0.06 (.07)	0.06 (.07)
Public sector (1 = yes)			−0.89* (.19)	−0.88* (.19)	−0.88* (.19)
F	5.36*	5.68*	10.56*	9.74*	8.87*
R^2	.01*	.02*	.04*	.06*	.07*
ΔR^2		.01*		.02*	.01*

Note. N = 763 without controls; 743 with controls. PW = Person-Workgroup. JS = Job Satisfaction. Unstandardized coefficients are reported. Standard errors are in parentheses.

*$p < .05$.

TABLE 5
Slope Estimates of Job Satisfaction on Union Participation at Low and High Levels of PW Fit

	Without Controls	With Controls
Low PW Fit (−1 SD)	−0.14*	−0.16*
	(.04)	(.04)
High PW Fit (+1 SD)	−0.01	−0.05
	(.04)	(.04)

Note. See notes for Table 4.

FIGURE 5
Job Satisfaction by Person-Workgroup Fit on Union Participation (shading represents 95% confidence intervals)

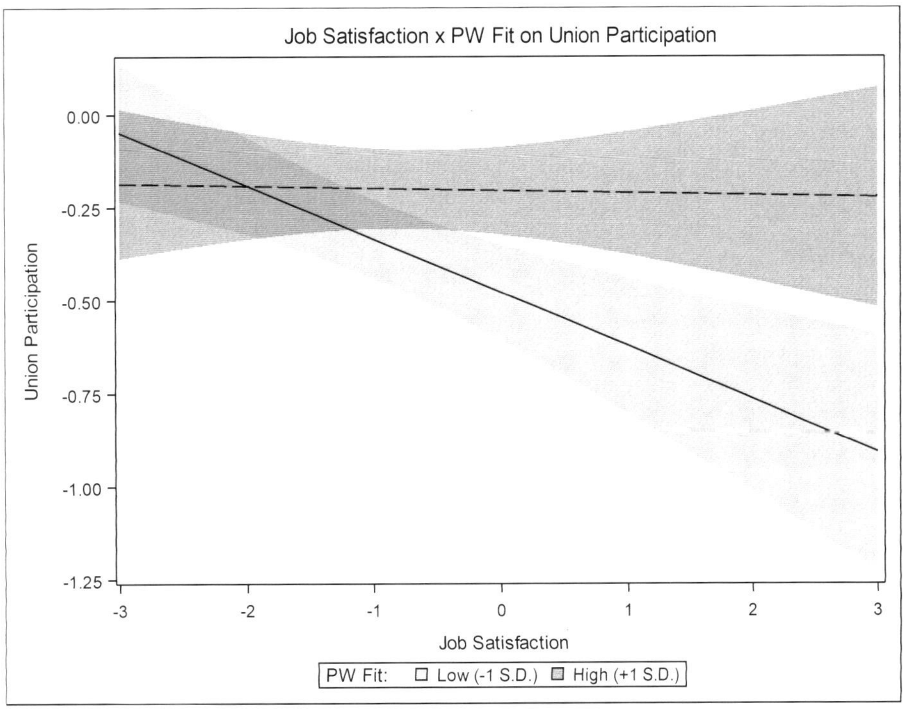

Discussion and Conclusions

Across five different workgroups in three labor unions covering both public and private sectors, results support fit's moderating role of job satisfaction and union participation. Specifically, a negative job satisfaction-union participation relationship exists only among low-fit workers. Job satisfaction and union participation's nonsignificant meta-analytic corrected correlation and wide credibility interval provided empirical impetus to unpack the oft-suggested relationship between the two variables. Our hypotheses followed consideration of the exit-voice tradeoff's application to union-only samples and the role of person-environment fit's relationship with participation. Following the attraction-selection-attrition framework, we proposed that high-fit individuals participate in unions regardless of job satisfaction because camaraderie and solidarity form more naturally than among poor fit workers. On the other hand, low-fit workers that do not experience satisfaction are relatively likely to either leave their employer or exercise voice by participating in their union to promote change. Low-fit individuals who are satisfied, but have no natural link to their coworkers stand out as a group less likely to participate—they have neither dissatisfaction nor workgroup fit to drive their participation.

Our results inform model development of union participation antecedents. Failing to account for moderators of the job satisfaction-union participation relationship can lead to erroneous conclusions. For example, if we had only evaluated job satisfaction and union participation broadly, we would have observed a degree of support for the exit-voice tradeoff based on the negative correlation between job satisfaction and union participation ($r = -.11$, $p < .05$). Yet, conditioning this relationship on fit rendered the relationship insignificant when fit was high and stronger when fit was low ($r = -.16$, $p < .05$). Thus, as evidenced by past meta-analyses, a negative job satisfaction-union participation relationship does not appear descriptive of all unionized workers. The bounds of job satisfaction's role as an antecedent to participation should be reconsidered conditioning on workgroup fit.

Unions may benefit from focusing organizing efforts on reaching workers with low fit and low job satisfaction. Our finding supports the idea that unions offer a direct "voice" alternative to leaving an employer on account of low job satisfaction *and* poor work fit. This corresponds with and extends Borjas's (1979) original union application of the exit-voice tradeoff. He suggested discrepancies between desired and actual work circumstances effectively activate the exit-voice tradeoff. High-fit individuals lack these person-environment discrepancies, but low-fit individuals do not.

The lack of a negative job satisfaction-union participation relationship for high-fit individuals sheds a positive light on union participation. If a negative job satisfaction-union participation relationship universally existed for workers, as suggested by the exit-voice tradeoff, the role of unions appears rather bleak: individuals who would otherwise leave their employer due to dissatisfaction can instead stay due to the voice their participation affords them, but their satisfaction remains low. Our findings, however, suggest such universal application of the exit-voice tradeoff is overly broad. Indeed, high-fit individuals participate with their union at relatively high levels *without* feeling dissatisfaction. Indeed, identifiable conditions exist where members are both satisfied with their work and engaged with their union. Unions and labor scholars should account for such boundary conditions.

Our study is not without limitations. Our sample is largely made up of public-sector union members, which may limit the generalizability of results. Nevertheless, since the public-sector union members were in a strike-permissive state, they arguably share means of participation more common to private-sector unions than public-sector unions in states not permitting strikes. We also controlled for the public- and private-sector distinction, and our findings are consistent with and without these controls. All considered, we suggest our results hold relevance for both public-sector unions in strike-permissive states and private-sector union members.

We are also limited in our ability to make causal claims regarding the effect of job satisfaction or fit on union participation due to the cross-sectional nature of our sample. Future studies might examine our hypotheses using longitudinal data. A more systematic examination of fit's moderation of the exit-voice tradeoff would include attrition data as well. An event history analysis similar to that of Iverson and Currivan (2003) that accounts for fit would thereby better serve to further clarify fit's role in the exit-voice tradeoff.

Future studies would also do well to examine the job satisfaction-fit interaction alongside other antecedents of union participation. Past model development benefited from the strong statistical power characteristic of meta-analyses (e.g., Bamberger et al., 1999; Monnot et al., 2011), but examinations of union participation and fit are rare and not likely to benefit from meta-analysis until more studies occur. Finally, future studies should consider whether fit distinguishes between job satisfaction differences in union and non-union samples. It may be that low-fit workers enabled by unions to stay on the job fully account for mean differences in job satisfaction between these groups. If this were the case, job satisfaction for high-fit workers would be the same for union and non-union workers—again providing a more hopeful take on the role unions play in the lives of their members.

In conclusion, this article expanded and linked two literatures not often studied together—labor relations and PE fit. Considering both led to proposing fit's moderating role of job satisfaction and union participation, which we examined using a primary sample of three diverse metropolitan unions. Our findings established boundary conditions to a relationship generally assumed negative. Our study helps guide the development of union participation models in future studies and informs the strategic choices unions face in organizing and reinvigorating lay activism. Unions stand to increase participation and lay activism by engaging low-fit, satisfied workers. Unions should also take heart in the distinction high fit affords—participation can occur alongside members' satisfaction just as easily as their dissatisfaction.

Acknowledgments

The authors gratefully acknowledge Fritz Drasgow and Ryan Lamare for their helpful reviews of earlier drafts of this article. Any remaining errors are our own.

Appendix: Developed Scales

Job Satisfaction in Union Settings Scale (α .88)

Indicate agreement with the following statements (1 = "Strongly Disagree" to 7 = "Strongly Agree," with "Neither Agree Nor Disagree" at Midpoint):

My job allows and encourages me to be creative and innovative.

Since starting my job, I have less time to do the job I was hired to do. (R)

Since being hired, my employer has made it easier for me to do my job.

I have become more fearful of maintaining my job. (R)

Since being hired, my employer has made my job more stressful. (R)

Given the way that my job is changing, I am confident that I will want to do this job for a long time.

While I want to keep doing this job, I am unhappy with the changes that my employer has implemented. (R)

N = 773; (R) = Reverse-coded item.

Overall Union Participation Scale (α .82)

Indicate participation relative to other employees in your union (1 = "Extremely Below Average" to 7 = "Extremely Above Average," with "An Average Amount" as midpoint):

I help with union organizing efforts.

I participate in union activities.

I have been/would be involved in strike support.

I have/would participate in picketing.

I volunteer for union service.

Indicate agreement with the following statements (1 = "Strongly Disagree" to 7 = "Strongly Agree," with "Neither Agree Nor Disagree" at Midpoint):

I would never engage in violence during a strike.* (R)

Picket line violence would not be justified even if factory management uses outside employees (scabs) to break a strike.* (R)

I would be willing to participate in an act of nonviolent civil disobedience in support of fellow workers, even if it meant that I could be arrested.

Under no circumstances would I support a work stoppage that was potentially illegal. (R)

N = 760; (R) = Reverse-coded item; *Original Militancy items from Martin (1986).

References

Aiken, L. S., and West, S. G. (1991). *Multiple regression: Testing and interpreting interactions.* Thousand Oaks, CA: SAGE Publications.

Anderson, J. C. (1979). Local union participation: A re-examination. *Industrial Relations, 18*(1), 18–31.

Bamberger, P. A., Kluger, A. N., and Suchard, R. (1999). The antecedents and consequences of union commitment: A meta-analysis. *The Academy of Management Journal, 42*(3), 304–318. https://doi.org/10.2307/256921

Borg, I., and Groenen, P. J. F. (2005). *Modern multidimensional scaling: Theory and applications* (2nd ed.). New York, NY: Springer,.

Borjas, G. J. (1979). Job satisfaction, wages, and unions. *The Journal of Human Resources, 14* (1), 21–40. https://doi.org/10.2307/145536

Brayfield, A. H., and Rothe, H. F. (1951). An index of job satisfaction. *Journal of Applied Psychology, 35*(5), 307–311. https://doi.org/10.1037/h0055617

Brown, S. D., and Gore, P. A. (1994). An evaluation of interest congruence indices: Distribution characteristics and measurement properties. *Journal of Vocational Behavior, 45*(3), 310–327. https://doi.org/10.1006/jvbe.1994.1038

Camp, C. C., and Chartrand, J. M. (1992). A comparison and evaluation of interest congruence indices. *Journal of Vocational Behavior, 41*(2), 162–182. https://doi.org/10.1016/0001-8791(92)90018-U

Fiorito, J., Padavic, I., and Russell, Z. A. (2014). Union beliefs and activism: A research note. *Journal of Labor Research, 35*(4), 346–357. https://doi.org/10.1007/s12122-014-9186-4

Freeman, R. B. (1980). The exit-voice tradeoff in the labor market: Unionism, job tenure, quits, and separations. *Quarterly Journal of Economics, 94* (4), 643–673.

Gordon, M. E., Beauvais, L. L., and Ladd, R. T. (1984). The job satisfaction and union commitment of unionized engineers. *Industrial and Labor Relations Review, 37*(3), 359–370. https://doi.org/10.2307/2523731

Hickey, R. (2005). Running your local. In J. Slaughter (Ed.), *A troublemaker's handbook 2* (pp. 290–305). Detroit, MI: Labor Notes.

Holland, J. L. (1959). A theory of vocational choice. *Journal of Counseling Psychology, 6*(1), 35–45. https://doi.org/10.1037/h0040767

Holland, J. L. (1963). Explorations of a theory of vocational choice and achievement: A four-year prediction study. *Psychological Reports, 12*(2), 547–594. https://doi.org/10.2466/pr0.1963.12.2.547

Holland, J. L. (1966). *The psychology of vocational choice.* Waltham, MA: Blaisdell Publishing Company.

Holland, J. L. (1997). *Making vocational choices: A theory of vocational personalities and work environments* (3rd ed.). Odessa, FL: Psychological Assessment Resources.

Huszczo, G. E. (1983). Attitudinal and behavioral variables related to participation in union activities. *Journal of Labor Research, 4* (3), 289–297.

Ironson, G. H., Smith, P. C., Brannick, M. T., Gibson, W. M., and Paul, K. B. (1989). Construction of a Job in General Scale: a comparison of global, composite, and specific measures. *Journal of Applied Psychology, 74* (2), 193.

Is TFA for you? (2016, July 18). http://bit.ly/2mRY3qH

Iverson, R. D., and Currivan, D. B. (2003). Union participation, job satisfaction, and employee turnover: An event-history analysis of the exit-voice hypothesis. *Industrial Relations: A Journal of Economy and Society, 42*(1), 101–105. https://doi.org/10.1111/1468-232X.00279

Judge, T. A., Hulin, C. L., and Dalal, R. S. (2012). Job satisfaction and job affect. In S. W. J. Kozlowski (Ed.), *The Oxford handbook of organizational psychology* (1st ed., Vol. 1). New York, NY: Oxford University Press.

Klandermans, B. (1986). Psychology and trade union participation: Joining, acting, quitting. *Journal of Occupational Psychology, 59*(3), 189–204. https://doi.org/10.1111/j.2044-8325.1986.tb00224.x

Kolchin, M. G., and Hyclak, T. (1984). Participation of union activities: a multivariate analysis. *Journal of Labor Research*, *5*(3), 255–262.

Kristof-Brown, A. L., and Guay, R. P. (2011). Person–environment fit. In S. Zedeck (Ed.), *APA Handbook of Industrial and Organizational Psychology* (Vol. 3, pp. 3–50). Washington, DC: American Psychological Association.

Kristof-Brown, A. L., Zimmerman, R. D., and Johnson, E. C. (2005). Consequences of individuals' fit at work: A meta-analysis of person–job, person–organization, person–group, and person–supervisor fit. *Personnel Psychology*, *58*(2), 281–342. https://doi.org/10.1111/j.1744-6570.2005.00672.x

Locke, E. A. (1969). What is job satisfaction? *Organizational Behavior and Human Performance*, *4* (4), 309–336. https://doi.org/10.1016/0030-5073(69)90013-0

Martin, J. E. (1986). Predictors of individual propensity to strike. *Industrial and Labor Relations Review*, *39*(2), 214–227. https://doi.org/10.2307/2523460

McShane, S. L. (1986). The multidimensionality of union participation. *Journal of Occupational Psychology*, *59*(3), 177–187. https://doi.org/10.1111/j.2044-8325.1986.tb00223.x

Monnot, M. J., Wagner, S., and Beehr, T. A. (2011). A contingency model of union commitment and participation: Meta-analysis of the antecedents of militant and nonmilitant activities. *Journal of Organizational Behavior*, *32*(8), 1127–1146. https://doi.org/10.1002/job.736

Nye, C. D., Su, R., Rounds, J., and Drasgow, F. (2012). Vocational interests and performance: A quantitative summary of over 60 years of research. *Perspectives on Psychological Science*, *7*(4), 384–403. https://doi.org/10.1177/1745691612449021

Parks, J. M., Gallagher, D. G., and Fullagar, C. J. A. (1995). Operationalizing the outcomes of union commitment: The dimensionality of participation. *Journal of Organizational Behavior*, *16*(7), 533–555.

Peterson, N. G., Mumford, M. D., Borman, W. C., Jeanneret, P. R., and Fleishman, E. A. (Eds.). (1999). *An occupational information system for the 21st century: The development of O*NET*. Washington, DC: American Psychological Association. http://bit.ly/2mNcJaK

Podsakoff, P. M., MacKenzie, S. B., Lee, J.-Y., and Podsakoff, N. P. (2003). Common method biases in behavioral research: A critical review of the literature and recommended remedies. *Journal of Applied Psychology*, *88*(5), 879–903. https://doi.org/10.1037/0021-9010.88.5.879

Prediger, D. J. (1982). Dimensions underlying Holland's hexagon: Missing link between interests and occupations? *Journal of Vocational Behavior*, *21*(3), 259–287. https://doi.org/10.1016/0001-8791(82)90036-7

Rounds, J., Su, R., Lewis, P., and Rivkin, D. (2010). *O*NET interest profiler short form psychometric characteristics: Summary*. Washington, DC: The National Center for O*NET Development.

Russell, S. S., Spitzmüller, C., Lin, L. F., Stanton, J. M., Smith, P. C., and Ironson, G. H. (2004). Shorter can also be better: The Abridged Job in General Scale. *Educational and Psychological Measurement*, *64* (5), 878–893. https://doi.org/10.1177/0013164404264841

SAS. (2012). (Version 9.4) [Windows]. Cary, NC: SAS Institute Inc.

Schneider, B. (1987). The people make the place. *Personnel Psychology*, *40*(3), 437–453.

Spector, P. E. (1997). *Job satisfaction: Application, assessment, causes, and consequences*. Thousand Oaks, CA: SAGE Publications.

Su, R. (2012). *The power of vocational interests and interest congruence in predicting career success* (Dissertation). University of Illinois, Urbana-Champaign. http://hdl.handle.net/2142/34329

Tracey, T. J. G., Allen, J., and Robbins, S. B. (2012). Moderation of the relation between person–environment congruence and academic success: Environmental constraint, personal flexibility and method. *Journal of Vocational Behavior*, *80*(1), 38–49. https://doi.org/10.1016/j.jvb.2011.03.005

Tracey, T. J. G., and Rounds, J. B. (1993). Evaluating Holland's and Gati's vocational-interest models: A structural meta-analysis. *Psychological Bulletin*, *113*(2), 229–246. https://doi.org/10.1037/0033-2909.113.2.229

Wheeler, H. N. (1985). *Industrial conflict: An integrative theory*. Columbia, S.C.: University of South Carolina Press.

Zener, T. B., and Schnuelle, L. (1976). Effects of the self-directed search on high school students. *Journal of Counseling Psychology*, *23*(4), 353–359. https://doi.org/10.1037/0022-0167.23.4.353

Building Internal Organisational Capacity to Respond Strategically to Neoliberalism: Insights from a Teacher Union in Australia

MIHAJLA GAVIN
The University of Sydney Business School

This paper examines how one teacher union in the state of New South Wales, Australia, the New South Wales Teachers' Federation (NSWTF), has built organisational capacity across its organisational, financial and governance structures since the 1980s to respond to neoliberal reform affecting collective organisation and the teaching profession. A qualitative research methodology is used with data drawn from union documents and interviews with 61 NSWTF officers and rank-and-file activists. Through applying David Weil's (2005) model of union strategic choice and decision-making, this study finds that renewal and revitalisation of historically and culturally embedded internal practices and structures in the NSWTF is critical to ensure the union adapts to present and future neoliberal challenges, that members can continue to engage in active, participatory democracy and decision-making at a grassroots level, and that credibility of the union is maintained in the eyes of the membership. This research contributes to debates and understandings of how teacher unions may deploy tactics and strategies to respond to neoliberal education reform and contemporary challenges facing unions.

Introduction

Since the early 1980s, neoliberal governments across Western democracies have sought to inject neoliberal and market ideologies into education policy in an effort to foster more competitive and productive national economies, which has subsequently transformed the work and conditions of teachers at a national as well as global level (see Connell 2013). Simultaneously, neoliberal governments have also endeavoured to undermine the strength and influence of the collective organisations that represent workers in the teaching profession—teacher trade unions—in a climate of general hostility towards the labour movement. Interestingly, despite declining trade union membership levels across unions on an international scale, teacher unions have been effective in retaining a comparative degree of strength and influence as measured by levels of union membership and density (Carter, Stevenson, and Passy 2010). Despite this phenomenon, little in known in existing literature about contemporary efforts being made by teacher unions to challenge or resist neoliberal education reform in order to assert the interests of their members and the teaching profession overall (Carter, Stevenson, and Passy 2010; Bangs and MacBeath 2012). Moreover, literature on union renewal and revitalisation has emphasised that unions within this climate need to consider how they can transform their internal structures and external activities to effectively adapt to contemporary challenges facing the labour movement.

This paper investigates the efforts of one teacher union in the state of New South Wales (NSW), Australia—the New South Wales Teachers' Federation (NSWTF)—to respond to neoliberal challenges affecting unionism and the teaching profession. The NSWTF is an industrial and professional union that represents the interests of teachers employed in the public education sector in NSW. In the face of neoliberal

Author's address: H70, Abercrombie Street and Codrington Street, Darlington, NSW 2006, Australia challenges, the NSWTF has been effective in surviving as an industrial organisation that can influence education policy and in retaining a high level of union membership and density. In particular, this paper will examine whether and how the NSWTF has built internal organisational capacity in a climate of neoliberal reform since the 1980s to adapt and revitalise its internal structures and processes in a way that, at its core, provides the necessary foundations to simultaneously resist neoliberal agendas whilst advancing the interests of its membership.

To achieve this end, this paper applies David Weil's (2005) model of strategic choice and decision-making to understand how the NSWTF has built organisational capacity across its organisational, financial and governance structures since the 1980s to respond to neoliberal challenges. It utilises a qualitative research methodology incorporating a case study approach with data sourced from internal and external union documents and 61 interviews with current and former NSWTF officers and rank-and-file activists. It will interrogate issues including whether and how the NSWTF has built strong internal foundations that facilitate membership-driven decision-making around the strategic direction and positioning of the union, how union renewal may be achieved across its organisational, financial and governance structures, and the extent to which teacher unions can overall "push back" and resist within a climate of hostility towards unionism.

Literature Review

This section will provide a brief overview of how neoliberal and marketised agendas have transformed education and teaching, the place of teacher unions within this context, and how teacher unions can strategically position themselves to respond to these agendas. It will also present the case study organisation used in this research.

Global Education Reform Through a Neoliberal Agenda

Education reform policies underscored by neoliberal and marketised agendas have transformed education systems and the work and conditions of teachers since the early 1980s. Neoliberalism is an economic philosophy and theory that advocates for economic and social transformation where domestic and global economic relations are restructured around market principles in the belief that such agendas will improve human well-being by encouraging individual choice and freedom (Olssen and Peters 2005). Neoliberalism represents a shift away from Keynesian concerns for social welfare and instead favours deregulating markets, liberalising trade, privatising sectors of the economy, creating "smaller" governments and promoting anti-union, "flexible" labour policies (Peetz and Bailey 2011).

Such ideas were popularised by right-wing economic and political theorists in the US and Europe and developed further through public choice theories adopted by the Thatcher and Reagan Governments in the 1980s (Edwards, Cahill, and Stilwell 2012). Across Western democracies, the reform of education has gained traction in the belief that restructuring schooling will generate economic success (Helsby 1999). Under neoliberal and marketised agendas, education policies have been reoriented from a liberal-humanist tradition where education is viewed as having intrinsic value to the individual and society to an instrumental approach where education is commodified as an untapped economic resource (Helsby 1999). The free market underpins neoliberal education reform discourse, with the idea that expanding parental choice will increase competition amongst schools, squeeze "poorer" schools out of the market, and improve the quality of education and student learning outcomes (Helsby 1999). In conjunction, the drive for greater competition has seen the rise of high-stakes testing and strict accountability practices led at both national and international levels (Rizvi and Lingard 2010).

Within this climate, the work and conditions of teachers have been restructured with significant consequences for teaching practice, professionalism, identity and collegial relations. Standardised testing has produced a narrowing of the curriculum and an emphasis on "teaching to the test," which has reduced choice and professional autonomy for teachers (Au 2011). Neoliberalism has also fostered a "technical-managerial" identity where the technical and instrumental aspects of teaching are advanced (Angus 2013) and "contrived collegiality" has seen a depersonalisation and humanisation of the student-teacher relationship (Hargreaves 1994).

Neoliberal Effects on Unions and Capacity to Respond

Along with neoliberalism affecting education and the work and conditions of teachers, neoliberalism has also affected the collective organisations representing the teaching profession. Neoliberal forces have encouraged growing employer and governmental assault on trade unions through legislative and regulatory changes that restrict traditional union rights and place threats on the wages and conditions of the workers represented by such unions (Hurd 2003). Public sector unions, such as teacher unions, have faced threats of privatisation, the expanded power of right-wing politics, restrictions on traditional union activity such as strikes, and limitations in negotiating with fiscally conservative governments (Hurd 2003).

Teacher unions have also been viewed as "standing in the way" of achieving flexibilities in the employment conditions and organisation of teachers' labour (Robertson 2000). Debate exists around the extent to which unions can "push back" against neoliberal forces and exert influence in a climate of declining union strength and influence. Two schools of thought exist on this issue. One side of this debate argues that unions are rarely compelled to act proactively and will only respond during a crisis (Stratton Devine and Reshef 1996) and may even ignore long-term threats in favour of short-term "satisficing" to maintain membership support (Ross and Martin 2006). The alternative side of the debate argues that union leaders should and often do respond proactively to changes in their external environment, but even this is carried out only in times when a union is not significantly constrained by external factors (Stratton Devine and Reshef 1996).

Strength of Teacher Unions

In considering how teacher unions have built organisational capacity within a climate of threats to unionism and education, it should be acknowledged that teacher unions have retained a considerable degree of strength and influence in representing the industrial and professional interests of their members. Compared with unions in the private sector, public sector unions (and particularly teacher unions) have retained high levels of membership and density. In Australia in the mid-1970s, over 50% of the workforce belonged to a trade union in their main paid jobs, dropping to 40% in the early 1990s and 30% by the mid-1990s (Sadler and Fagan 2004: 31). Latest figures from August 2013 show only 17% of all employees across Australia are union members in their main job (ABS 2013).

By comparison, teachers on a global level have remained a highly unionised sector of workers despite neoliberal forces threatening collective organisation (Carter, Stevenson, and Passy 2010). Teachers are the largest group of knowledge workers associated with collective labour (Kerchner, Koppich, and Weeres 1997) as reflected by statistics which show that in Australia, employees in the education and training industry had the highest proportion of union membership based on latest data from 2013 (ABS 2013). This trend is also reflected in other Western democracies where the education industry had the second highest rate of union membership in 2015 in the US (Bureau of Labor Statistics 2016) and the highest rate in 2014 in the UK (Department for Business Innovation and Skills 2015).

Union Strategy and Building Organisational Capacity

Although teacher unions have retained a comparative degree of strength within a climate of threats to unionism, it is important to consider whether and how teacher unions have attempted to respond strategically and engage in practices of renewal and revitalisation to survive and thrive as collective organisations. In a time of declining union membership, diminishing power resources and declining capacities to mobilise, labour movements have begun to confront some of these issues by revitalising and renewing their internal and external operations and activities in order to remain relevant and influential within a turbulent environment facing union (Cornfield and McCammon 2003; Frege and Kelly 2003; Hyman 2007). This study will consider ways that teacher unions have reorganised their internal practices and processes in order to build the foundations for active, engaged and participative forms of unionism (Fairbrother 2000).

In conjunction with understanding how teacher unions have engaged in internal renewal and revitalisation practices, it is important to consider whether and how unions have engaged in such practices strategically. While the term "strategy" has been defined in many and varied ways, most definitions consider strategy as a deliberate and conscious set of guidelines that determine decisions into the future (Mintzberg

1979). Strategy does not occur in singular forums but manifests in the ongoing internal matters of an organisation and as part of a union's general activities which mutually reinforce to shape its direction (Weil 2005). While Kochan, Katz and McKersie's seminal text of 1986 introduced the concept of strategic choice to the industrial relations literature, since this time, the conceptualisation of union strategic management and choice in the literature has generally been slow (Frege and Kelly 2003) with little insight into whether and how unions have used strategy to address and combat key challenges (Stratton Devine and Reshef 1996). Given the numerous environmental challenges facing unions today, unions need to effectively adapt and display proficiency making strategic decisions (Stratton Devine and Reshef 1996).

David Weil's model of strategic choice and planning provides a systematic framework that unions can use to make strategic choices that adapt to challenges in their environmental context (Weil 2005). Weil argues that the strategic situation facing a union can be considered along two dimensions: strategic level and organisational capacity (Weil 2005). "Strategic leverage" reflects the external environment in which unions operate including labour market challenges, technological factors, work organisation, and legislative and policy frameworks. "Organisational capacity" considers the union's internal organisational, financial, and governance structures (Weil 2005). A union generally has greater ability to affect its organisational capacity than its strategic leverage by translating key strategic decisions into activities that are undertaken by the union in the desired places (Weil 2005). By developing both strategic leverage and organisational capacity, unions can adapt to external challenges and increase the likelihood achieving long-term objectives (Weil 2005). Each element of "organisational capacity" is considered below.

Organisational Structure

Unions can build organisational capacity by adjusting their organisational structure. A union's organisational structure includes the formal organisation such as the size of the union, different levels and divisions and authority figures, and also the people working within that structure including elected officers and appointed staff (Weil 1994). It also consists of the human resource system which consists of methods for recruiting, selecting, training, promoting and compensating staff, the overall culture of the union such as the organisational mission, management and leadership style, and how information is shared within the union (Weil 1994). However, while unions may adopt new strategies, progressing these strategies is often hindered by old structures and organisational legacies (Bramble 2001; Weil 2005). Unions are typically viewed as 'path dependent" and will not threaten shared ideals and habits to which they have long been accustomed (Hyman 2007). Moreover, as unions grow they inevitably develop bureaucratic, centralised structures and experience an ongoing tension between hierarchical control and democratic decision-making (Boxall and Haynes 1991). According to Hyman (2007), union organisational structures should be considered a 'process' where established routines and assumptions can be 'unlearned' when they are no longer effective.

Financial Structure

A union's 'fiscal health' can greatly affect the strategic choices available. 'Fiscal health' is determined by sources of revenue, size of expenses relative to revenue, and the surplus or debt that accumulates over time (Weil 1994). 'Fiscal health' is maintained when a union can forecast levels of expenditure and sustain chosen strategies despite changes in external conditions (Weil 1994). A union should also consider how "dues-dependent" it is and move towards a position of establishing a stable financial base that is less reliant on member dues (Weil 1994). Additionally, unions need to consider how money is collected and distributed within the union, who authorises expenditure, and decision-making processes around allocating financial resources (Weil 1994).

Governance Structure

The governance structure of a union is generally established by its constitution and articulates the democratic practice around how decisions are made and policy is formed and implemented (Weil 1994). A union's governance structure includes its formal structure, people and culture. The formal governance structure determines where major policy issues develop and who has authority for implementing those policies. A union typically comprises of a large decision-making body consisting of representation from local union branches which acts as the highest policy-making body, as well as smaller governance bodies that meet more regularly to consider the implementation of policies (Weil 1994). The degree of centralisation or decentralisation in a union's governance structure is usually determined by the union's constitution (Weil 1994). Unions should also consider the culture of decision-making, which can be strongly influenced by the union's leadership (Weil 1994).In line with literature on union renewal, unions need to consider how to rebuild workplace strength through encouraging greater rank-and-file involvement in decision-making (Cornfield and McCammon 2003). Additionally, as unions grow, emphasis should be placed on decentralising governance structures and increasing activities at the local branch level to minimise concerns of unions becoming too large or unwieldly (Spaull and Hince 1986).

Case Study of the New South Wales Teachers' Federation

This study explores strategic decisions around the organisational capacity of one teacher union, the New South Wales Teachers' Federation. The NSWTF is the trade union in the state of New South Wales, Australia, representing teachers in the public school system. It represents teachers employed in pre-school, infants, primary and secondary schools, as well as teachers working in vocational education and training and corrective service centres (NSWTF Annual Report 2015: i). The NSW education system comprises a mix of public (government), Catholic and private (non-government) schools with 65.1% of students attending public schools in NSW (ABS 2015). Compared to other states and territories, NSW also teaches the largest number of students (ABS 2015).The NSWTF formed in 1918 and is the largest industrial union in NSW with a total membership in 2015 of approximately 73% of all full-time, permanent teachers across the state (NSWTF Annual Report 2015: 86).

At present, the NSWTF employs 46 full-time union officers in various senior leadership and professional roles such as industrial/research, welfare support, field organisers as well as specialised coordinator positions overseeing areas including Aboriginal issues, women and multiculturalism. The NSWTF considers itself as both an industrial and a professional union concerned with the industrial interests and working conditions of its members and in seeking improvements for public education more generally (Zadkovich 1999; Fitzgerald 2011). At the NSWTF's core is a focus on democracy and participation of members in its decision-making processes. The union's relationship with its members and its strong internal culture are considered key reasons for the union's "survival" in a climate of hostility towards unionism and public education in NSW (Fitzgerald 2011).

Methodology

A qualitative research methodology utilising a case study approach has been used to understand how the NSWTF has built organisational capacity. Constructionist and interpretivist methods are suitable when conducting exploratory research which can lead to hypothesis building and explanations (Ghauri and Gronhaug 2010). This study has sought to investigate how the NSWTF has built organisational capacity to support union renewal and build collective strength. The time period for this study spans the mid-1980s to present, reflecting consensus in the literature around the time of emerging neoliberal agendas. A single case study approach has been adopted to identify factors involved in the functioning and behaviour of the NSWTF (Ghauri and Gronhaug 2010). Although the number of observations from a single case study may be low and reduce capacity for generalisability, the emphasis in qualitative research is to better understand and examine the social world as interpreted by its participants. Providing "thick description" enables understanding of the particularities and complexities of the subject under investigation (Geertz 1973; Ghauri and Gronhaug 2010).

Data was collected through document analysis and semi-structured interviews. Documents of the NSWTF that were analysed include annual reports, decisions at annual conference meetings, minutes, decisions at council and executive meetings, and articles from the union's journal. Documents can provide historical insight about the organisation and enhances validity as there is less researcher obtrusiveness and reactivity (Bowen 2009).

To complement insights from document analysis, 61 interviews were conducted with current and former union officials and activists of the NSWTF, comprising 30 women and 31 men. As this research spans a 30 year period, it was necessary to capture the perspectives of participants who were both currently active (20 participants) and formerly active (41 participants) within the NSWTF across a diverse range of leadership and grassroots positions. To preserve participant confidentiality, the researcher has not specified current or former engagement. As this research concerned understanding internal processes of the NSWTF, most participants are union officials who can provide deep insights into these processes. However, insights were also garnered from long-serving rank-and-file activists who could reflect on the NSWTF's decision-making processes. In many cases, the interviewed activists also held roles as councillors or as members of the executive. Semi-structured interviews were used to deeply investigate phenomena and "probe" answers to questions (Yin 2011), which is important when gauging the perceptions, experiences and thoughts of participants (Saunders, Lewis, and Thornhill 2007).

Table 1 outlines the positions held by participants at the time of interviewing. To preserve anonymity, the age, sex, or whether the position is currently or formerly held has been omitted.

TABLE 1
Positions Held By Participants at Time of Interview

Position	Total	Identifier
President	5	President (P) 1; P2; P3; P4; P5
Deputy President	5	Deputy President (DP) 1; DP2; DP3; DP4; DP5
Senior Vice-President	2	Senior Vice-President (Senior VP) 1; Senior VP 2
General Secretary	7	General Secretary (GS) 1; GS2; GS3; GS4; GS5; GS6; GS7
Deputy Secretary/Assistant General Secretary	5	DP/AGS 1; DP/AGS 2; DP/AGS 3; DP/AGS 4; DP/AGS 5
City Organiser	7	City Organiser (CI) 1; CI2, CI3, CI4, CI5, CI6, CI7
Country Organiser	9	Country Organiser (CO) 1; CO2; CO3; CO4; CO5; CO6; CO7; CO8; CO9
Other Organiser	1	Other Organiser 1
Women's Coordinator	3	Women's Coordinator (WC) 1; WC2; WC3
Research/Industrial Officer	3	R/I Officer (R/I) 1; R/I Officer 2; R/I Officer 3
Professional Support Officer	3	PS Officer (PS) 1; PS Officer 2; PS Officer 3
NSWTF Journal Editor	1	Editor 1
Trade Union Training Officer	1	Training Officer 1
Activist and/or Executive Member/Councillor	9	Activist 1; Activist 2; Activist 3; Activist 4; Activist 5; Activist 6; Activist 7; Activist 8; Activist 9

Documents were first analysed by searching out underlying themes and organising information into categories related to the central question of understanding the NSWTF's organisational capacity (Bowen 2009). Interview questions were developed from the themes arising out of document analysis and through discussions with key informants and a comprehensive review of the relevant literature. An interview schedule and questions were developed which explored themes and issues around the NSWTF's organisational, financial and governance structure as informed by Weil's framework on union strategy. Interview data and documents were coded and analysed using qualitative data analysis software NVivo around themes and sub-themes emerging from interview responses and analysis of documents (Glaser and Strauss 1967).

Findings

Organisational Structure

This section will examine three key components of the NSWTF's organisational structure over the last 30 years—how officers are hired, succession and training for officers, and officer tenure—and will consider where organisational capacity has and can be built in order to strengthen the union within a climate of threats to unionism and the education industry.

Hiring Officers—Elections, Officer Background, Skills, and Expertise

Senior officers of the NSWTF are elected by the membership every 2 years and all other officers are elected every 3 years and subject to re-election by ballot of the council (NSWTF Annual Report 2013: 11). This practice of election is also observable in other parts of the NSWTF's structure including election of delegates to the council and annual conference by associations, election of executive members by councillors, and election of the local union representative in each school (Other Organiser 1; PS3; CO2). The process of electing officers for positions by the rank-and-file membership rather than appointment by application was viewed as a highly democratic practice that created a "sense of ownership" (Other Organiser 1) and authenticity (DP/AGS2; CO4; CO2; PS3; Activist 4).

Some criticisms, however, did exist around this process. Participants commented that being reasonably well-known was essential to be elected, typically meaning that nominees would "pander to the electorate" (Other Organiser 1; R/I2). There was also a view that councillors would avoid voting for an officer who was "totally ideologically out of step" (Other Organiser 1) or thought "outside the box" (DP4), sometimes creating a perception that the union was a "closed shop" (Activist 9), which challenged notions of democracy and the ability to bring different perspectives or ideas into the union (Other Organiser 1; R/I2; PS2). Criticism also surrounded the fact that in recent years fewer officers were losing their positions through challenge by other candidates, and officers (particularly senior officers) were often elected unopposed, potentially due to lack of interest in standing (Other Organiser 1; R/I2; CO2; CO8).

The NSWTF also has a strong tradition of only electing officers from within the teaching profession. This practice was a "sacred cow" (CO3) and a measure of strength and credibility for the union where officers came from within the profession and were represented by rank-and-file teachers (GS3; CO5; CO7; CI5; CI7; Other Organiser 1; R/I3; WC3). This practice allowed officers to understand the experience of "being at the chalkface" (WC3) when engaging with members and having currency of experience positively influenced interactions with members (GS3; CI1; CI3; CI5; CO4; CO8; R/I2). Additionally, there is a view that since officers are elected from within the profession, decision-makers would actively consider impacts on students and public education in their decision-making (Senior VP1; CI3; CI5; CO7). This style of hiring was also favoured in comparison to the vast majority of other unions that hire professional union officers who may not have prior practical experience in the relevant industry (Senior VP1; CO7; CO8; Other Organiser 1; R/I2; R/I3; WC3).

However, some participants also expressed that while officers have experience in teaching, there were no other skills, competencies or expertise required of candidates to stand for nomination, and hiring decisions were exclusively based on popular election (DP/AGS2; R/I2). Some participants also expressed concern that it was not a requirement to have experience in leadership or management or training in managing budgets, which were critical skills for senior leadership positions (GS1; PS2). In this vein, the NSWTF was criticised as a "closed" organisation that did not seek external advice or expertise in hesitation that such practices may "open the flood gates" to more of this practice (DP/AGS2; CI1; CO3; Activist 7; Activist 9). Some also questioned the level of experience officers had in their teaching career prior to becoming an officer. There was a view that the NSWTF was unfavourably developing a culture where officers have only a few years of teaching experience prior to becoming an officer (CI2; CI7; CO1; CO7; R/I2). Participants felt that officers needed to first establish themselves as teachers and gain credibility and experience across different types of employment in order to speak to members with a level of authority (CO3; R/I1).

Succession Planning and Training

Participants held different views about how succession planning and promotion occurred in the NSWTF. There was general consensus that formalised succession planning was "anathema to [union's] culture" (P5) and was driven by the "politics of the [internal] factions" (DP/AGS1) where senior officers would encourage "like-minded people" and "their own favoured candidates" (P5) during elections (GS5; R/I1; PS2). However, a view was also shared that although promotion of officers could be faction-driven, that over time, leadership in the union had come from different factional strands (P5) and that individuals who were "exemplary" (R/I1) would generally be mentored and supported in the formation of the next cohort of experienced officers (P5; GS1; GS5; DP/AGS2).

Participants also held differing views about the extent and quality of induction and ongoing training for officers. Most observed that training generally lacked rigor and formalisation. In more recent years, new officers were able to attend a three-day training course run by the Australian Education Union (the national teachers' union in Australia) (CI3; CO4) and also participate in training provided by other unions such as the Australian Council of Trade Unions (peak body for trade unions in Australia) or other universities or regulatory bodies (CI3; Training Officer 1; R/I1). The provision of a budget for training also provided some support for officers with identified areas of development (Training Officer 1).

Generally, training that was available for officers was ad hoc and informal. An important informal training platform was provided when officers would meet at Federation's Head Office prior to council meetings and have the opportunity to discuss different issues with the general secretary (P5; GS1; CI6; Training Officer 1). It was also common for officers to receive training from their colleagues or via on-the-job experience, which was assisted by enhancements in technology that facilitated improved communications between officers and a strong internal culture that fostered good collegial support (GS5; DP/AGS4; CI1; CI2; CI4; CI5; CO9; R/I2). This was particularly critical for organisers who are geographically dispersed across the state (DP/AGS4; CI6). Officers could also use some of their prior skills and knowledge as a teacher activist or local union representative when taking up an officer position (GS5; CI7; CO4; R/I2) or have the opportunity to engage in temporary relief work when an officer in their substantive position was on leave (CO5; CO9).

Many participants were quite critical about the lack of formalised training that was available to new officers. Many highlighted how the union's belief in a "common sense" approach to training (P5; CI5; CI7; R/I2) based on "assumed knowledge" (CI4) and "prior skills" (DP/AGS1) made officers feel they were being "thrown in" to the organisation (CI5; CO4) without proper induction or mentoring (GS1; CO5; CO4; CO 5; CI3; CI7) and expected to "sink or swim" (DP/AGS2; R/I2; R/I3). As expressed by one officer: "[lack of training] has significant pitfalls because you go from, at best, you're a Fed Rep [local union representative] at a school, to going out there and people ringing you up on such serious matters" (CI4).

Officer Tenure

Participants were sharply divided about whether limited tenure for officers should be introduced. Whilst this concept was floated as early as 1986, a policy around limited tenure has not been introduced by the union (NSWTF 1986 "Review of Federation Structure"). Participants in favour of limited tenure for officers observed how there were fewer officers returning to teaching after completing their 3 year elected term, which had seen a rise of long-term professional officers who had built a career in the union (CI5). To be a "good officer" and an effective activist, participants believed that it was essential for officers to understand the daily experiences of teachers at the classroom level and have currency of teaching experience (P4; CO5; WC2; PS3; Activist 2). In senior leadership positions, limited tenure was viewed, particularly by activists who were interviewed, as essential as greater time out of the teaching profession could mean "los[ing] some perspective on the day-to-day dealings" (CI2; R/I1; Activist 4; Activist 6; Activist 9). Problematically, the union is also currently in a position where many long-serving senior officers will soon be retiring, threatening retention of corporate and cultural knowledge of the organisation (GS1).

By contrast, many participants rejected notions of limited tenure. Some participants commented that, to a considerable extent, officers were not actually union officials but rather "deployed teachers working for the union" (P5) on a temporary basis who are subject to re-election and only provided with 6 years of leave from their employer (the NSW Department of Education) to undertake external activities (DP2; CO5; R/I3). Some felt that particular positions in the union, such as organiser or welfare roles, allowed officers to still be in "tuned in to local issues" (CI5; R/I3; PS3; Activist 5). Limited tenure was also not favourable given that the complexity of the industrial and political environment meant that officers were still learning about the external environment (DP/AGS1; CO2; Other Organiser 1). Over time, officers also built up considerable knowledge and diverse skills of value to the union (GS6; DP/AGS2; CO3; CO9), particularly in senior leadership roles where knowledge, expertise and political astuteness was crafted over decades of working for the union (CI2; CO1; CO5; CO9; Activist 4; Activist 6).

Participants generally favoured establishing a process that enabled officers to return to the classroom regularly in order to maintain a connection to the grassroots level. Although practicalities in establishing this process were acknowledged, allowing officers to regularly return to the classroom would provide officers with new perspectives and a greater connection with students and teachers, as well as foster improved trust, credibility and respect amongst the membership (CI1; CO1; CO7; R/I1; PS2; Activist 8). Some current officers even described how being out of the teaching profession for a long time made them feel "deskilled" as a teacher (GS6; CI4; R/I2).

Financial Structure

This section examines the financial processes and practices of the NSWTF and considers areas where the union has built organisational capacity in its financial management and decision-making in recent decades and further areas for consideration if the NSWTF is to continue operating with a sound financial base.

Revenue Sources and Income

In a climate of declining union membership and threats to trade unionism, the NSWTF has needed to consider its various revenue sources. Pertinent to this issue is the question of the extent to which the NSWTF is "dues-dependent." According to Weil's framework, most unions depend substantially on member dues for revenue and while some dependence is inevitable, the relative level of dependency can affect the strategic position of the union (Weil 1994). Weil suggests that unions should move towards a position of establishing a stable financial base where they are less "dues-dependent" and focus more on other revenue sources (Weil 1994). For the NSWTF, its "lifeblood" has been membership fees (Senior VP1). During the 1980s and 1990s, membership fees accounted for close to 90% of the NSWTF's income, however, since this time has consistently exceeded 90% (Senior VP1; GS1; GS2; G4; GS6; DP/AGS3; PS3; CI1; Activist 4; NSWTF 2015 Annual Report: 28; AEU NSWTF 2016: 7).

To a lesser degree, the NSWTF also operates other income streams. During the 1990s, the union considered it important to begin diversifying its income stream to avoid "[putting] all our eyes in one basket" (DP2; GS6; GS7). This saw the union purchase and lease other floors and neighbouring retail and commercial businesses of NSWTF's Head Office building to secure rental income (DP2; GS5; GS6; GS7). However, while such income supplies several million dollars annually, it is exclusively viewed as "icing on the cake" (GS1). Another important income stream is the Public Education Fund Levy established in the late 1990s, where a percentage of membership fees are allocated to an exclusive campaign fund (CO8; DP/AGS3). The campaign fund can be spent to fund campaigns centred on public education issues, which has given the NSWTF the capacity to "lead the debate" (CI4) when it comes to issues facing students' interests and public education more broadly (CI4; GS5).

There were differing perspectives amongst participants around whether or not membership fees being the main source of income was considered problematic. Although some participants acknowledged membership fees was a critical source of income for the NSWTF, others noted that it could be problematic if this was the *only* source of income and potential fluctuations in membership levels could affect the union's service offerings (GS6; PS2; CI9; CO6; Activist 7). However, by and large, this source of income was viewed as critical to the NSWTF's "independence" (DP2) and "strength" (CI4). Participants agreed that if the union began to rely on other non-union related sources of income that it could challenge this independence and incur a "political cost" (DP2; CI4). Reliance on membership fees for income not only represented a "monetary measure" but also a "symbolic measure" of the level of member engagement in and commitment to the union's campaigns and commitment (CI4). Incidentally, there was an obvious need for the NSWTF to maintain its high membership level and ensure that recruitment was an active activity engaged in by organisers and councillors in their interactions with teachers (GS1; CO9). This is particularly critical given challenges currently being faced by the NSWTF in the changing demographic profile and employment structures of the teaching profession, notably the retirement of significant numbers of "baby boomer" teachers and the increased casualisation of the teaching workforce in NSW (DP/AGS4; CI5; CO3; CO4; CO5; R/I2; PS2; Activist 6; NSWTF Annual Conference Decisions 2004: 13; NSWTF 2009 "Everyone's a winner with a strong union": 11).

Financial Management and Reporting

Key processes have been implemented in the NSWTF to ensure effective financial management and reporting of finances. The NSWTF operates a Finance Committee consisting of executive members, a senior officer, and the NSWTF's Accountant, which meets at least monthly to discuss finances in consultation with the general secretary (GS1; GS2; GS4; GS5; CI7). Various forums exist where the NSWTF's General Secretary is required to report on the state of the union's finances (GS4) including publication of accounts from the previous year at the annual conference and a detailed reporting of the union's proposed budget for the forthcoming year to the council (GS5; GS6; DP/AGS3). Since the early 1980s, the NSWTF has also employed a full-time accountant to work alongside the general secretary (GS2; GS6). While employing an accountant sparked a "hot debate" given the NSWTF's long-standing culture of only employing staff from the teaching profession, as the size of the union grew it became necessary to employ that particular skills set within the union (P4).

The NSWTF also prided itself on having a thorough attitude towards the transparency and accountability of its finances and considered itself as a leader for other unions in this area. Participants acknowledged that particularly in the last decade, the NSWTF had become "very strict about money" (GS6) so as to avoid attracting criticism over its financial activities (DP3; GS6). Having "checks and balances" over financial decisions, such as ensuring dual agreement over financial decisions, was considered imperative to avoid potential allegations of financial corruption (GS6; CI7; CO7). Advances in technology had also enabled the NSWTF to confidently "track every cent" of money coming into and out of the union (GS6). The NSWTF also saw itself as a financially conservative organisation. As to be explained below, whilst the NSWTF experienced previous financial difficulties, attitudes adopted by its general secretaries during the mid-1990s and 2000s saw the NSWTF adopt a more conservative attitude towards spending (DP1; GS5; GS6). Through being parsimonious in its spending, the NSWTF has been able to enjoy financial security and stability in recent decades (DP3; GS5; DP/AGS3).

Learning from Financial Difficulties

Some practices of NSWTF's financial management have improved in light of previous financial difficulties faced by the union. Many participants reflected on times in the union's recent history where deductions of member fees at source was halted by the NSW Department of Education. Traditionally, fees were deducted from members' salaries by the employer and paid to the union via cheque (P4; DP1; GS4; GS6; GS7). However, in 1974 and 1994, the respective NSW governments of the day withdrew this service following industrial disputes (P4; Senior VP1; GS1; GS7; CO2; CO3), resulting in a "financial crisis" (P4; CO3) which "nearly crippled" the union (J. Dixon) on both occasions. However, the shift away from deductions at source has been a slow process. It was not until 2006 that the NSWTF took proactive steps to initiate an ongoing campaign of encouraging members to deposit fee money via direct debit (GS1; DP1; DP3; Senior VP1; CO8; Activist 4). Currently, approximately 80% of members now use the direct debit method (GS1). The NSWTF's General Secretary has commented that such practice has attracted the attention of other teacher unions in the US that are experiencing similar challenges in the deduction of fees (GS1). In addition to encouraging members to change their method of payment, the NSWTF also established a special fund where a percentage of membership fees was allocated for "rainy day" emergency situations (GS6) in the event that the NSW government decided again to stop deductions in source, ensuring that the NSWTF had sufficient operating funds for a number of months (GS2; GS6; DP/AGS3; CO3; CO8).

Allocation of Finances and Budgetary Decision-Making

Budgets are set in the NSWTF by the general secretary seeking submissions from various sections, drawing up the budget, and presenting it to the union's Finance Committee and the executive for review, and finally to the council as a recommendation for debate and endorsement (P1; GS4; GS5). Regarding the allocation of finances, while the annual conference sets the broad categories of expenditure, participants noted that the general secretary and senior officers hold supreme decision-making power in terms of budgetary decisions, as informed by the recommendations and decisions of the council and the executive. For the NSWTF, the biggest area of financial expenditure was officer and staff salaries, accounting for over 50% of spending (GS5; GS6; NSWTF 1995 :How Federation spends your fees") with the remainder typically spent on campaign activities, travel expenses to attend various decision-making forums, and other legal, administrative, and affiliation fees (GS5; CO6; CO8; NSWTF 1995 "How Federation spends your fees"). It was important for the NSWTF to track spending on salaries given the significant financial outlay and consider where flexibilities could be achieved in the replacement of retiring staff (GS6; CO3).

As stated, the NSWTF generally considers itself to be a financially conservative organisation and a view was held that the union should consistently have at the forefront of its decision-making thinking that spending of union money was spending of members' money (GS5). When allocating money to campaigns, there was an understanding that the union needed to work within its resources and be "realistic" (CO4; Activist 4) about financial decisions given there wasn't a "bottomless pit" of financial resources (GS6; DP/AGS3; CO4; PS3; Activist 4). General Secretary's commented that the union would not go into a campaign that was not pre-funded and having a balanced budget was critical in the event of falling membership levels (GS2; GS6). There was also capacity, however, for the NSWTF to re-allocate money within the budget to invest more in campaign activities if a campaign intensified or a new campaign emerged (GS5; GS6; CI7). Given that the NSWTF had been running campaigns for so long, there was a view that the union confidently knew how much certain activities would cost and could therefore allocate funds appropriately, and had also adopted more sophisticated decision-making in understanding what activities would be most cost-effective when engaging in campaigns (CO4; Activist 4).

Governance Structure

This section will examine the governance structures that have allowed the NSWTF to be a union founded on democratic decision-making and will consider avenues to build further organisational capacity into the future to ensure that the membership continues to drive decision-making and steer the union through its present challenges.

Decision-Making Bodies

Three main decision-making bodies guide decision-making in the NSWTF—the annual conference, the council, and the executive. The NSWTF's annual conference brings together some 600 rank-and-file members each year to debate the union's policy positions and determine campaigns for the forthcoming year (DP4; DP/AG2). Decisions set at the annual conference automatically override council and executive decisions (DP/AG2; CI3). Tactics and strategies decided in this forum are then workshopped by the council in meetings held eight times per year attended by 300 rank-and-file teacher activists (known as councillors) (DP3; GS4; DP/AGS3; DP/AGS4; CO1; CO4; R/I1; Activist 1; Activist 4; Fitzgerald 2011). The council is one of the NSWTF's most important and active forums and its level of representation is higher than many other governmental bodies in Australia (DP/AGS4; CO1; CO4). Councillors from different teaching backgrounds and geographical areas can come together to discuss issues (GS6; CO3) and also directly pose questions to the union's senior officers (DP/AGS3). Ideas presented by senior officers require the endorsement of the council to proceed (GS6; NSWTF 1994 Council Minutes) and are then reported back to schools and associations by councillors or organisers (GS3; GS6; DP/AGS3; CI5).

The third major tier of decision-making is the executive, which comprises the union's senior officers and 15 practising classroom teachers who meet on a fortnightly basis to discuss tactics and issues arising between council meetings (Fitzgerald 2011). The executive can challenge recommendations from senior officers as well as propose recommendations for the council and can also be convened quickly in the event that critical interim decisions need to be made (GS3; CO1; CO4). The NSWTF also operates geographically situated associations where monthly meetings are held to discuss local issues facing teachers and schools.

Areas of Change and Resistance to Altering the Decision-Making Bodies

Participants considered the degree to which the NSWTF's main decision-making bodies had changed since the 1980s to meet the evolving needs of the union. A notable example was reducing the size of the union's annual conference in the late 1980s from over 1000 delegates to around 600-700 delegates. This action was taken following views from the membership at the time that such a large forum hindered its effectiveness as a decision-making body because it was becoming unwieldly and extremely costly to operate (DP4; GS5; Other Organiser 1; CO3; PS3). The number of council meetings held annually was also reduced to eight per year for cost-effectiveness (GS5; CI1; CO3; NSWTF 1990 Executive Minutes). A suggestion to reduce the number of councils further in the mid-2000s was narrowly defeated (DP2; GS6; CI2; CI4; NSWTF 2005 "Facing the future in a hostile environment") and similarly, the proposal to have a bi-annual instead of annual conference was also defeated (DP/AGS2). Participants generally felt that such decisions were necessary to improve the cost-effective running of the NSWTF and did not appear to diminish the level of democracy and ability for members to participate in union decision-making.

Whilst there have been some notable changes to the size of the NSWTF's decision-making bodies, by and large these structures have remained much the same since their establishment in the 1950s (DP/AGS5; CI2; R/I2; NSWTF Annual Conference Decisions 1999). Most participants saw these governance structures as working effectively and as delivering democratic decision-making, and whilst from time-to-time there have been discussions around potential further changes, no serious conversations have occurred to this effect. To a lesser extent, there has been active resistance and reluctance to change these structures (CO3; Activist 8). It is interesting to note also that the NSWTF opted against continually transforming its structure in comparison to the employer, the NSW Department of Education, which seemingly has been undertaking a constant process of internal restructuring (DP5; NSWTF Executive Minutes 1991).

Capacity for Member Voice in Decision-Making Bodies

This study investigated the extent to which these main decision-making forums provide a genuine opportunity for members to participate in and shape the strategic direction of the union. The council was considered by many participants to be "one of the most powerful debating chambers in the country" (Senior VP1) where robust debates saw no single view being shared and where decision-making was not confined to a small group of people (DP1; CI3; CO3; CO9). If dissatisfied, councillors had the capacity to vote down recommendations from the senior officers or executive (Senior VP1; GS3; CO1), which helped to "keep [the senior officers] on our toes" and "not be complacent" (GS3). This level of debate and capacity to put forward well thought out arguments and reasoning also meant that ideas and strategies were thoroughly "tested almost to destruction by 20 other people trying to attack it, as opposed to simply saying this is the president's view, rubber-stamp it." (DP1; Senior VP1; CI6; CO9). However, some participants also felt that in comparison to council debates of the 1970s and 1980s, there was less intensity of debate nowadays and there was more "guided democracy" (Senior VP1; GS6; R/I2; PS2). Additionally, given that recommendations brought by senior officers were generally comprehensive meant that, to a large degree, some "rubber stamping" of recommendations still occurred (GS3; GS6; R/I2).

In addition, this research also considered the degree to which decision-making was driven more by the leadership or membership of the union. The majority of participants stated that decision-making was driven both top-down and bottom-up depending on what issue required decision-making (P1; P4; GS4; GS7; Senior VP1; CI1; CO1; CO6; CO7; Activist 1; Activist 4). Many participants shared the view that the role of the senior officers was to provide analysis to the membership who can then provide input into implementation of strategy. Whilst the strategic direction of campaigns was generally established by the leadership, members had considerable involvement at the grassroots levels and from their associations in being able to implement decisions on the ground appropriate for their area and circumstances (P1; DP3 DP4; GS7; CO4; CO5; Activist 2). There was a view that a sign of good democracy in the union was senior officers not readily getting out of step with membership sentiments, typically meaning that leadership needed to play close attention to the views of the membership (DP4; CO1; Activist 5).

Some participants were more critical in expressing that decision-making was generally driven "top-down" by senior officers who play a significant role in determining what campaigns will be run and how they will play out (DP1; DP4; CI3; CI5; CI7; CO8; PS2; Activist 4; Activist 8). At the grassroots level, there was also some pressure felt by the council to accept the decisions made by senior officers (Activist 7). Participants also observed some issues that could affect the level of representation and democracy at council meetings. Regarding the composition of members elected to attend the council forum, more awareness needed to be fostered by the union in the late 1990s and early 2000s to ensure that the composition of its decision-making forums was keeping pace with changes in the demographic profile of the teaching profession in NSW. Union literature in the early 2000s discussed how with the average age of teachers in NSW reaching their late 40s, the NSWTF's decision-making forums had begun to reflect this ageing process (NSWTF 2000 "New activists encouraged"). Therefore, more consciousness needed to be built in ensuring that the council is representative in all sections of its membership, including younger teachers and women (NSWTF 1999 Annual Conference Decisions).

Discussion and Conclusion

An analysis of the NSWTF's organisational capacity and ways the NSWTF has built capacity since the 1980s demonstrates that the union has established strong foundations to ensure that the collective voice of the teaching profession is being represented, that members can drive decision-making to positively affect the strategic positioning of the NSWTF in a context of threats to collective organisation and struggles in advancing the interests of teachers amidst neoliberal education reform, and that a strong financial structure can support its campaign activities into the future. However, in line with the literature on renewal and revitalisation, there are several key areas where the NSWTF can potentially challenge some historically and culturally embedded internal practices in order to adapt to changing external conditions and ensure that it is consistently engaging with its members at a grassroots level, encouraging informed decision-making, and maintaining credibility in the eyes of its members.

Although there are limitations in generalising from a single case study, there is capacity from these findings to reveal insights for other teacher unions of the specific practices that have successfully served the NSWTF as a long-standing and influential union in NSW, and contributed to the overall favourable strategic positioning of teacher unions within the current labour movement. The findings do not intend to imply a direct cause-and-effect between how the organisational capacity of the NSWTF has measurably impacted its effectiveness as a union or comment on its external activities, but rather have sought to convey the perceptions of informed respondents about the internal practices and processes that have, and can be, adopted to remain a strong and influential union in a climate of threats by neoliberal agendas affecting unionism and education.

Regarding the NSWTF's organisational structure, the NSWTF had well established hiring practices to ensure that officers of the union were effectively representing the teaching profession on behalf of their peers. However, for the union to effectively represent its members, the findings reveal that officers needed to have credibility in the eyes of the membership and be better equipped in various ways to maintain that credibility. In some cases, the "organisational legacies" of the NSWTF hindered officers from being able to establish true credibility in the eyes of members (Bramble 2001; Weil 2005).

Having officers elected by the rank-and-file membership and hiring officers only from within the teaching profession represented hallmarks of the strength and credibility of the NSWTF. However, given that officers generally were elected based on their aligned views and attitudes of the organisation, which potentially created complacency in members standing for election, there may be opportunity for the union to consider, where relevant, engaging external consultants or experts from the education field on a short-term basis to provide advice and expertise on education-related issues to ensure other voices of authority can positively shape the strategic direction of the union, while maintaining credibility in the eyes of the membership by having an officer core derived from the teaching profession.

Additionally, while participants overwhelmingly commented that having teachers elected from within the profession was important for the sake of representation and ensuring that officers understood issues facing teachers when making decisions, rules around officer tenure meant there was less opportunity for officers to return to the classroom and understand the daily issues facing teachers "at the chalkface." Participants observed that officers needed to have currency of experience in order to speak to members with credibility and authority. Whilst implementing rules around tenure and number of years' experience before standing as an officer would be unlikely and unfavourable, there is potential for the union to facilitate a process of officers re-engaging with teachers at the classroom level at more regular periods, particularly officers whom may have limited daily contact with teachers in their roles. While implementing tenure rules for senior officers would be particularly unfavourable given their breadth of expertise and corporate knowledge, as senior officers readily shape the strategic direction of the union, facilitating a process of re-engagement would be critical at this level.

Although having a formalised succession planning process was not favoured by the union and was generally incongruent with the election process, there was a critical need for more formal and rigorous training for officers. Although in recent years, new officers were able to engage in some formal induction training, overwhelmingly, officers felt that more formalised and regular training was needed at the introductory officer level and when advancing into more senior roles, given the unique skills set required of an officer and the complexity of member issues being handled. This was particularly important given that no specific skills or competencies, particularly around leadership or financial management, were required for prospective officers.

Analysis of the NSWTF's financial structure since the 1980s showed that the NSWTF has adopted more financial savviness over the decades in understanding how and where its money is spent and in taking proactive steps to safeguard the flow of money into the union. However, further safeguards are also required if the NSWTF is to remain a financially sound union into the future. While Weil's framework emphasises that unions should move towards a position of establishing a stable financial base where they are less dues-dependent (Weil 1994), overwhelmingly, participants in this study viewed members' dues being the main source of income as an important marker of the NSWTF's strength and independence, which was unlikely to change. It is important, however, in light of changing external conditions such as the shifting demographic profile and employment structures of the teaching profession in NSW, that the NSWTF continues to consider how recruitment can be embedded into the daily work of its officers and activists if it desires to remain an organisation in good "fiscal health" (Weil 1994).

Regarding its formal financial structure, participants reflected positively on the centralised and decentralised ways in which it was determined how money was collected and distributed within the union (Weil 1994). Although senior officers generally had supreme decision-making over the allocation of finances, financial decision-making was also informed by the views of members at council and executive meetings (Weil 1994). Having open and transparent processes for financial management allowed the NSWTF to avoid "mystique," typically observed in the literature, around internal union accounting whereby financial data is often the sole domain of accounts either subcontracted by the union or working in siloed finance departments (Weil 1994).

It was clear from participants that sophisticated and transparent processes had been adopted by the union's leaders over time to consistently and rigorously analyse resource allocation and ensure that members also had oversight of financial decisions about the allocation of resources during a budget period (Weil 1994; Weil 2005). As reflected in Weil's model, the NSWTF demonstrated proficiency in being able to sustain campaign strategies despite changes in external conditions as well as gauge and forecast its level of expenditure for a fiscal period in order to establish plans within the context of those expected expenses (Weil 1994). The NSWTF also effectively considered resources in its decision-making processes around allocating money for campaign activities and favoured adopting a financially conservative approach to spending members' money and managing the size of its expenses relative to revenue in order to maintain good "fiscal health" (Weil 1994). Evidence of being able to strategically re-evaluate how member fees were collected by the union following several financial crises (whilst somewhat reactive) was also demonstrated by the NSWTF, showing ability for the union to adapt entrenched financial practices and systems in order to meet changing external conditions (Weil 1994).

The governance structures in place in the NSWTF allowed the membership to drive the strategic direction of union decision-making. Various layers of decision-making in the union reflected consideration of how decisions are made and policy is formed and implemented (Weil 1994). Consideration was also given to the degree to which decision-making is centralised or decentralised (Weil 1994). It was clear that a decentralised governance structure was in place whereby senior officers could not unilaterally make decisions removed from the membership and needed to secure the endorsement of members around decisions on policy direction and implementation. It was also clear that it was practising teachers embedded in the daily experience of teachers at large who were informing the strategic direction of the union.

Although some change had occurred in the size of the union's decision-making structures, the overall governance structure continued to facilitate representation and input of members and it was not seen that democracy was lost through this process. Only moderate observations were made of resistance to changing the structures to which the union had long been accustomed (Hyman 2007). This conveys some evidence of the governance structure in the NSWTF being a "process" and not a rigid structure, where there was potential to adapt to changing needs and circumstances of the union (Hyman 2007).

Further, there did not appear to be overt tensions between hierarchical control and democratic decision-making in the NSWTF (Boxall and Haynes 1991). Generally, decision-making was driven equally by the leadership and membership, depending on the issue, and officers and activists acknowledged the important role played by senior officers in setting the strategic direction of the union and ensuring members had input into guiding that strategic direction through the governance structure. There was only some suggestion that decision-making was driven more top-down by senior officers. Additionally, there was moderate criticism that although the council was a vigorous debating forum where ideas could be tested and good policy was formed, in more recent years the council appeared less active and rigorous. In light of these observations, it is important that teacher unions, such as the NSWTF, encourage open, active, participatory debating forums where dissenting views of members can be expressed and there is a balance of centralised and decentralised decision-making to ensure that members continue to guide decision-making in the union. This is important for union renewal in considerations of how unions can encourage greater rank-and-file involvement in decision-making (Cornfield and McCammon 2003).

Similarly, as part of the union renewal process, there is evidence that changes in the demographic profile of the NSWTF's membership necessitate the union to consider how to ensure the full cross-section of its membership is being represented and can access participatory forums. With an ageing teaching profession in NSW and fewer young teachers attending the NSWTF's decision-making forums, there was evidence of disproportionate representation at council meetings, which can potentially affect the union's workplace strength and democracy if full representation is not achieved (Cornfield and McCammon 2005). There is therefore potential for the NSWTF to further consider how it might engage newer members in its decision-making forums to ensure full representation and input into decisions as well as a strategic direction shapes by the needs and concerns of contemporary teachers.

In conclusion, this study has found that whilst teacher unions, such as the NSWTF, have established historically strong internal foundations for members to drive democratic decision-making within the union and have core input into the strategic direction of the union, elements of these historically and culturally embedded internal structures may similarly constrain teacher unions from renewing and revitalising in the face of neoliberal challenges to unionism and the teaching profession. It is important that within a climate of continuing hostility towards unionism that teacher unions act proactively in considering where there is scope and potential to adapt their organisational, financial, and governance structures to ensure that members continue to shape the strategic direction and positioning of the union, credibility of the union is maintained in the eyes of the membership, and that the union retains solid yet flexible foundations on which to advocate for the interests of the teaching profession.

References

Australian Bureau of Statistics. (ABS) 2013. *Employee Earnings, Benefits and Trade Union Membership*, Australia, 6310.0, August 2013.

Australian Bureau of Statistics. (ABS) (2015) Schools, Australia, 2014. Cat. no. 4221.0. ABS, Canberra, ACT.

Australian Education Union New South Wales Teachers Federation Branch (AEU NSWTF). 2016. Financial Statements, 31 December 2016.

Angus, Lawrence. 2013. "Teacher Identity and the Neoliberal Condition: Asserting a Participative-Professional, Socially Democratic Teacher Imaginary in Technical-Managerial Times." *Journal of Educational and Social Research,* Vol. 3, no. 7, pp. 170-177.

Au, Wayne. 2011. "Teaching under the new Taylorism: high-stakes testing and the standardization of the 21st century curriculum." *Journal of Curriculum Studies*, Vol. 43, no. 1, pp. 25-45.

Ball, Stephen J. 2003. "The teacher's soul and the terrors of performativity." *Journal of Educational Policy*, Vol. 18, no. 2, pp. 215-228.

Bangs, John., and John MacBeath. 2012. "Collective leadership: the role of teacher unions in encouraging teachers to take the lead in their own learning and in teacher policy." *Professional Development in Education*, Vol. 38, no. 2, pp. 331-343.

Boxall, Peter., and Peter Haynes. 1997. "Strategy and Trade Union Effectiveness in a Neo-liberal Environment." *British Journal of Industrial Relations*, Vol. 35, iss. 4, pp. 567-591.

Bowen, Glenn A. 2009. "Document Analysis as a Qualitative Research Method." *Qualitative Research Journal*, Vol. 9, no. 2, pp. 27-40.

Bramble, Tom. 2001. "Australian Union Strategies since 1945." *Labour & Industry*, Vol. 11, no. 3, pp. 1-25.

Bureau of Labor Statistics. 2016. *Union Members Summary*, United State Department of Labor, USDL-16-0158, 28 January, 2016.

Carter, Bob., Stevenson, H., and Rowena Passy. 2010. *Industrial Relations in Education: Transforming the School Workforce.* New York: Routledge.

Connell, Raewyn. 2013. "The neoliberal cascade and education: an essay on the market agenda and its consequences." *Critical Studies in Education*, Vol. 54, no. 2, pp. 99-112.

Cornfield, Daniel B., and Holly J. McCammon. 2003. "Revitalizing Labor: Global Perspectives and a Research Agenda." In Daniel B. Cornfield and Holly J. McCammon, *Labor Revitalization: Global Perspectives and New Initiatives.* Oxford: Elsevier Ltd, pp. 1-20.

Department for Business Innovation and Skills. 2015. *Trade Union Membership 2014: Statistical Bulletin*, BIS/15/300, June 2015.

Edwards, Lindy., Cahill, Damien., and Frank Stilwell. 2012. "Introduction: understanding neoliberalism beyond the free market." In Lindy Edwards, Damien Cahill and Frank Stilwell, *Neoliberalism: Beyond the Free Market.* Massachusetts: Edward Elgar, pp. 1-12.

Fairbrother, P. 2000. *Trade Unions at the Crossroads.* London: Mansell Publishing.

Fitzgerald, Denis. 2011. *Teachers and Their Times: History and the Teachers Federation.* Sydney: University of NSW.

Frege, Carola M., and John Kelly. 2003. "Union Revitalization Strategies in Comparative Perspective." *European Journal of Industrial Relations*, Vol. 9, no. 1, pp. 7-24.

Gardner, Margaret. 1989. "Union Strategy: A Gap in Union Theory." In Bill Ford and David Plowman eds., Australian Unions 2nd edn. South Melbourne: MacMillan, pp. 49-72.

Geertz, Clifford. 1973. "Thick Descriptive: Toward an Interpretive Theory of Culture." In Clifford Geertz ed., *The Interpretation of Cultures: Selected Essays.* New York: Basic Books, pp. 3-30.

Ghauri, Pervez, and Kjell Gronhaug. 2010. *Research Methods in Business Studies, 4th edn.* England: Pearson.

Glaser, Barney G., and Anselm L. Strauss. 1967. *The Discovery of Grounded Theory: Strategies for Qualitative Research.* Chicago: Aldine Publishing Company.

Hargreaves, Andy. 1994. *Changing Teachers, Changing Times: Teachers' Work and Culture in the Postmodern Age.* London: Cassell.

Helsby, Gill. 1999. *Changing Teachers' Work.* Philadelphia: Open University Press.

Hurd, Richard W. 2003. "In Defense of Public Service: Union Strategy in Transition." *Working USA*, Vol. 7, no. 1, pp. 6-25.

Hyman, Richard. 2007. "How can trade unions act strategically?" *Transfer,* Vol. 13, no. 2, pp. 193-210.

Kerchner, Charles T., Koppich, Julia E., and Joseph G. Weeres. 1997. *United Mind Workers: Unions and Teaching in the Knowledge Society.* San Fransisco: Jossey-Bass Inc.

Kochan, Thomas A., Katz, Harry C., and Robert B. McKersie. 1986. *The Transformation of American Industrial Relations.* New York: Basic Books Inc.

Minztberg, Henry. 1979. "Patterns in Strategy Formation." *International Studies of Management and Organizations*, Vol. 9, no. 3, pp. 67-86.

New South Wales Teachers' Federation (NSWTF). 1986. "Review of Federation Structure." *Education.* 30 June, p. 16. NSW: NSWTF.

New South Wales Teachers' Federation (NSWTF). 1990. *1990 Executive Minutes, "Reducing the Size of Annual Conference and Council,"* 13 February. NSWTF.

New South Wales Teachers' Federation (NSWTF). 1991. *1991 Executive Minutes, "Federation Structures,"* 25 June. NSWTF.

New South Wales Teachers' Federation (NSWTF). 1994. *1994 Council Minutes, "Powers and Duties of Council,"* 28 May. NSWTF.

New South Wales Teachers' Federation (NSWTF). 1995. "How Federation spends your fees." *Education.* 22 May. NSW: NSWTF.

New South Wales Teachers' Federation (NSWTF). 1999. *1999 Annual Conference Decisions, "Challenges facing Teacher Unionism in Its Organisation."* NSWTF.

New South Wales Teachers' Federation (NSWTF). 2000. "New activists encourages." *Education.* 10 April. NSW: NSWTF.

New South Wales Teachers' Federation (NSWTF). 2004. *2004 Annual Conference Decisions, "Recruitment and activism and the future of the union."* NSWTF.

New South Wales Teachers' Federation (NSWTF). 2005. "Facing the future in a hostile environment." *Education.* 7 March. NSW: NSWTF.

New South Wales Teachers' Federation (NSWTF). 2009. "Everyone's a winner with a strong union." *Education.* 6 April. NSW: NSWTF. NSWTF. 2013. *2013 Annual Report.* Marrickville, NSW: NSWTF.

New South Wales Teachers' Federation (NSWTF). 2015. *2015 Annual Report.* Thornleigh, NSW: NSWTF.

Olssen, Mark., and Michael A. Peters. 2005. "Neoliberalism, higher education and the knowledge economy: from the free market to knowledge capitalism." *Journal of Education Policy*, Vol. 20, no. 3, pp. 313-345.

Peetz, David., and Janis Bailey. 2011. "Neoliberal Evolution and Union Responses in Australia." In Gregor Gall, Adrian Wilkinson and Richard Hurd eds., *International Handbook on Labour Unions: Responses to Neo-liberalism.* Cheltenham: Edward, Elgar, pp. 62-81.

Rizvi, Fazal., and Bob Lingard. 2010. *Globalizing Education Policy.* Oxon: Routledge.

Robertson, Susan L. 2000. *A Class Act: Changing Teachers' Work, Globalisation and the State.* New York: Falmer Press.

Ross, George., and Andrew Martin. 2006. "European Unions Face the Millennium." In Andrew Martin and George Ross eds., *The Brave New World of European Labor: European Trade Unions at the Millennium.* USA: Berghahn Books, pp. 1-25.

Sadler, David., and Bob Fagan. 2004. "Australian Trade Unions and the Politics of Scale: Reconstructing the Spatiality of Industrial Relations." *Economic Geography*, Vol. 80, no. 1, pp. 23-43.

Saunders, Mark., Lewis, Philip., and Adrian Thornhill. 2007. *Research Methods for Business Students, 4th edn.* England: Pearson.

Spaull, Andrew., and Kevin Hince. 1986. *Industrial Relations and State Education in Australia.* Blackburn: Australasian Education Press Pty Ltd.

Stratton Devine, Kay., and Yonatan Reshef. 1996. "Union planning: a framework and research agenda." *Industrial Relations*, Vol. 51, no. 3, pp. 506-523.

Voss, Kim., and Rachel Sherman. 2000. "Breaking the Iron Law of Oligarchy: Union Revitalization in the American Labor Movement." *American Journal of Sociology*, Vol. 106, no. 2, pp. 303-349.

Weil, David. 1994. *Turning the Tide: Strategic Planning for Labor Unions.* New York: Lexington Books.

Weil, David. 2005. "A Strategic Choice Framework for Union Decision Making." *WorkingUSA: The Journal of Labor and Society*, Vol. 8, no. 3, pp. 327-347.

Yin, Robert K. 2011. *Qualitative Research from Start to Finish.* New York: The Guilford Press.

Zadkovich, Gary. 1999. To Deliver Us Our Future: Developing Strategy for Difficult Times. NSW Teachers' Federation, Eric Pearson Report.

VI. Work Organization and Public Policy I

Agglomeration Economy and Competition Improve Productivity and Employment in Local Service Industries

SACHIKO KAZEKAMI
Chukyo University

I demonstrate that 1) the density of establishments in retail trade, wholesale, the narrowly defined service industries, and moreover, manufacturing increases the total factor productivity (TFP) in each service industry, 2) the agglomeration economy improves the TFP of retail trade; and 3) competition improves the TFP of wholesale and the narrowly defined service industries. As the Japanese society has been facing a decline in population, especially in the rural areas, I focus on rural municipalities and find that they have lower productivity in the service sector. However, the agglomeration economy improves the TFP of the service sector in these municipalities.

Introduction

Improving the productivity of the service sector is important for Japan as well as other developed countries because it accounts for a large share of their economies. In Japan, the value added by the service sector accounted for 72.9% of the GDP in 2015. If we limit the service sector to retail trade, wholesale, eating and drinking services, hotel, service professionals, technology and business services, education, health and welfare, and other services (i.e., the non-tradable service industries), the value added by these sectors accounted for 38.5% of the GDP, while the manufacturing sector accounted for 20.4% of the GDP. In Japan, employment in the service sector accounted for 72.3% of all workers in 2015. Employment in the limited service sectors described above accounted for 55.4% of all workers. Therefore, improving productivity is necessary for economic growth. Furthermore, it is necessary for increasing wages because productivity positively correlates with wages in most cases (Morikawa 2016).

Previous studies have sought the factors that improve the productivity of service sectors in competition, the externality effect of human resources, high operating rate, and rational production plan due to large demand, and management know-how of the manufacturing sector. Syverson (2004, 2011) argues that competition induces Darwinian selection or the improvement in efficiency within firms. Darwinian selection implies that competition transfers market share towards more efficient producers, reducing the number of relatively high cost firms/plants by forcing their exit at times and opening up room for more efficient producers. Heightened competition can also induce firms to undertake costly productivity-raising actions that they may otherwise avoid, leading to an increase in efficiency within plants or firms.

Syverson (2004) investigates the connection between competition and productivity in a case study of the ready-mixed concrete industry. He finds that the productivity distribution of ready-mixed concrete plants is truncated from below as the demand density rises. Markets with denser construction activities have higher lower-bound productivity levels, higher average productivity, and lesser productivity dispersion. In another study, Foster, Haltiwanger, and Krizan (2006) find that aggregate productivity growth in the U.S. retail sector is almost exclusively because of the exit of less efficient single-store firms and their replacement with more efficient national chain store affiliates.

Lucas (1988), Rauch (1991), and Moretti (2004) discuss the externality effect of human resources. Moretti (2004) estimates plant-level production functions using a unique firm-worker matched data set. He demonstrates that the spillovers in two industries located in the same city and are economically close are larger than the spillovers in two industries located in the same city, but are economically distant. Morikawa (2014) notes that the scale economy, high

Author's address: Department of Economics, Chukyo University, 101-2 Yagoto Honmachi, Showa-ku, Nagoya-shi, 466-866, Japan operating rate, and rational production plan due to large demand are factors that can improve the productivity level in addition to the above factors. Morikawa (2014) uses a survey of selected service industries and chooses specific industries, i.e., cinemas, golf courses, tennis clubs, fitness centers, bowling alleys, golf driving ranges, cultural centers, matrimonial agencies and wedding ceremony hall brokers, and aesthetic salon services, and finds that a doubled population density increases productivity in the service sector by 7-15%. Previous studies analyze the effect in the same service industries. Moreover, previous empirical studies examine specific service industries or limited service industries. I shed light on not only the effect of the service sector, but also on other sectors such as the manufacturing sector. I examine the non-tradable service industries that account for a large part of the GDP and employment, discussed in more detail below

Beyond academic studies, Japan Revitalization Strategy (2016), which is the plan that improved the productivity of the service sector in Japan, proposes methods such as using the know-how of the manufacturing sector in the service sector, consulting local financial institutions, making an information center for IT knowledge, listing the specialists who know new services, visualizing the quality of service, and certifying services.

There are company-related factors such as management know-how and market environment-related factors such as competition that can affect the productivity of the service industries. However, as policy-makers should consider the market environment-related factors, I analyze it in this study. Among market environment-related factors, I consider the agglomeration economy and competition.

The agglomeration economy yields a knowledge spillover effect, a skill spillover effect, and benefits from sharing labor and intermediate inputs markets. Furthermore, I consider two types of spillover effects, that is, the spillover effect within an industry and that from other industries. As described above, competition induces Darwinian selection or the improvement of efficiency within firms. I roughly analyze the mechanism at work after finding out the positive effects of competition on productivity.

In particular, I examine the municipalities that specialize in the health and nursing sector and analyze the effect of an agglomeration economy of the service and manufacturing sectors on the productivity of the service sector. There is a decrease in population in Japan and many municipalities do not expect a high density of establishments. The health and nursing care industry is the main source of job opportunities in some rural municipalities of Japan. However, it is a non-tradable service industry. Therefore, my analysis is focused on those municipalities.

The service sector consists of a variety of industries. I focus on the retail trade industry, wholesale industry, and the narrowly defined service industries (service professionals, technology and business services, education, health and welfare, and other services)[1] because a typical characteristic feature of the service sector is the non-tradable industry, and this induces a typical issue of the service. sector. A non-tradable industry is one in which demand is limited by the size of the local economy (e.g., volume of value added, residents' income level, and population size). Employment in wholesale and retail sales, which are representative non-tradable service industries, accounts for the largest share, 24%, of the total employment (10.54 million employees) in the service sector of Japan. Employment in the medical and welfare industry, which are the other representative non-tradable service industries, accounts for 17% of the total employment in the service sector. It is also the fastest growing industry in Japan. I measure productivity through total factor productivity (TFP) and labor productivity.

In the next section, I explain the empirical approach used in the study. In section 3, I describe the data. I present the estimation results in section 4. Finally, I discuss and conclude the paper in section 5.

Empirical Approach

First, I investigate the factors that improve the level of productivity in the service sector using the following equation:

$$Productivity_{jt} = \beta_0 + \beta_1 X_{jt} + \beta_2 Cont_{jt} + \beta_3 Year\ dummy_t + \mu_{jt} \quad (1)$$

$Productivity_{jt}$ denotes the log of TFP or the log of labor productivity. X_{jt} denotes the factors in municipality j in year t that improve the productivity of the service sector. I use the density of establishments in retail trade, wholesale, and the narrowly defined service industries, its square, and the log of the number of establishments in the service sectors to investigate the spillover effect in the same industries. I also use the density of establishments in manufacturing and the log of the number of establishments in the manufacturing sector to investigate their spillover effect. The density of establishments in the service sector and the log of the number of establishments in the service sector do not allow distinguishing between the effects of an agglomeration economy and competition. Therefore, I use the modified Ellison-Glaeser measure (Ellison and Glaeser 1997) for an agglomeration economy and the Herfindahl-Hirschman Index for competition. The Ellison-Glaeser measure is the standard measure of proxies for the presence of knowledge spillover, labor market pooling, input sharing, and so on (Pe'er, Vertinky, and Keil 2016). I do not include the share of employment in each firm. Therefore, I use the following measure:

$$EG_{jt} = \left(X_{jt} - I_{jt}\right)^2 \qquad (2)$$

X_{jt} denotes municipality *j*'s share of total employment in all industries and I_{jt} denotes municipality *j*'s share of employment in the retail trade, wholesale, or the narrowly defined service industries in year *t*. I obtain the aggregate data on employment in retail trade and wholesale industries. The. Herfindahl-Hirschman Index is calculated using the following equation:

$$C_{jt} = \Sigma_k S_{kjt}{}^2 \qquad (3)$$

S_{kjt} denotes the share of sales of firm *k* in retail trade, wholesale, or the narrowly defined service industries in year *t*. I estimate Equation 1 by using the fixed effect model, and cluster them according to municipality. I control economic change over time with the year dummy. Additionally, I control for the differences in educational level and the demography by region with $Cont_{jt}$. I use the ratio of people who graduate college/university or graduate school by prefecture level, and the ratio of people older than 65 years old in the residence by municipality. Moreover, I examine Equation 1 with and without income per resident or sales of each service industry per resident. Income per resident or sales per resident capture the (potential) level of demand. Income per resident is not income per worker, which indicates the level of productivity in the old theory. Income per worker may capture the part of effect of an agglomeration economy because it increases the productivity of workers. However, I capture the effect of an agglomeration economy by the variable *X*, rather than by income per worker.

To estimate the TFP, I estimate the production function using Levinsohn and Petrin's (2003) model, which is as follows:

$$y_{it} = \beta_0 + \beta_l l_{it} + \beta_k k_{it} + \beta_m m_{it} + \omega_{it} + e_{it} \qquad (4)$$

y_{it} denotes the log of output (sales) of firm *i* in year *t*, l_{it} denotes the log of number of workers in firm *i* in year *t*, kit denotes the log of capital of firm *i* in year *t*, mit denotes intermediate inputs of firm *i* in year *t*, and ω_{it} denotes unobservable productivity shocks. All firms are from retail trade, wholesale, or the narrowly defined service industries. The demand function for intermediate input is given as:

$$m_{it} = m_t(\omega_{it}, k_{it})$$

It must be monotonic in ω_{it} for all k_{it} to qualify as a valid proxy. I estimate this function by a two-step estimation method. From this production function, I calculate the log of TFP of firm *i* in year *t*, TFP_{it}. Then, I calculate the average log of TFP in city *j* and the average log of TFP during two years. To estimate the log of labor productivity, I divide sales in the retail trade and wholesale industries by the aggregate number of workers in retail trade and wholesales as I obtain only that data. Furthermore, I estimate the effect on the number of workers by using the number of workers as the dependent variable instead of productivity in Model 1.

Second, I investigate whether Darwinian selection or efficiency improvement works. I calculate the ratio of establishments that enter or exit the market to total establishments in each municipality.

Third, I focus on the municipalities where the health and nursing is the main industry because of the lack of other industries in some rural municipalities. I demonstrate that the level of productivity in the municipalities that specialize in health and nursing is lower than that in other municipalities. However, (1) the agglomeration of retail trade, wholesale, or the narrowly defined service industries, and (2) the agglomeration of manufacturing increases the productivity of the service industries. First, I estimate the following model:

$$Productivity_{jt} = \alpha_0 + \alpha_1 Health_{jt} + \alpha_2 density_{jt} + \alpha_3 density^2 + a_4 density \times Health_{jt} + \alpha_5 density^2 \times Health_{jt} + \alpha_6 Cont_{jt} + \varepsilon_{it} \quad (5)$$

$Health_{jt}$ denotes the health and nursing dummy in municipality j in year t; the health and nursing dummy is equal to one if a municipality has a specialized index[2] greater than one in the health and nursing industry, otherwise zero. $density_{jt}$ and $density^2$ are the density of establishments and their squares, respectively, in retail trade, wholesale, or the narrowly defined service industries. $density \times Health$ and $density^2 \times Health_{jt}$ are cross-terms between the health and nursing dummy and the density or its square. $Cont_{jt}$ is the difference in educational level and the demography of a region. I analyze whether α_1 is negative and α_4 is positive.

Second, I simply divide the municipalities into those with a specialized index greater than one and those without a specialized index greater than one in the health and nursing industry. Moreover, I divide the sample by the number of manufacturing industries that have a specialized index greater than one is one to six, or not. I demonstrate the evidence using the distribution chart. I analyze this evidence using econometric method too. The results are available upon request.

Additionally, I limit the sample to metropolitans and analyze the above estimations for. robustness. Moreover, I estimate the whole sample for commuting zones instead of municipalities for robustness. The reason is that economic activity often integrates across municipal borders. However, I do not include these robustness results because of a space constraint. The results are almost similar to the main results. These results are available upon request.

Data

I analyze the model using data from 1995, 2000, 2005, and 2010 as some data are not available for 2015. As for TFP, I use two years and calculate the average, of the years 1995, 1996, 2000, 2001, 2005, 2006, 2010, and 2011. I obtain data on sales, value added, tangible fixed assets, intermediate inputs (cost plus sales and general administration cost minus payroll, depreciation, welfare, rent, and tax) and employment from the Basic Survey of Japanese Business Structure and Activities conducted by the Ministry of Economy, Trade and Industry for the period from 1995 to 2011. The survey is administered to enterprises with 50 or more employees that have excess capital or investment funds valued at over 30 million yen. This survey collects data by each firm. Therefore, I use the Establishment and Enterprise Census to incorporate the information detailing the locations of the establishments in the Basic Survey of Japanese Business Structure and Activities.

The Establishment and Enterprise Census is conducted for all establishments in Japan by the. Ministry of Internal Affairs and Communications. I divide the data in the Basic Survey of Japanese Business Structure and Activities by the number of establishments and aggregate the data by municipality. As the municipalities are sometimes merged during the estimation periods, I adjust those cases. The Basic Survey of Japanese Business Structure and Activities does not cover small firms, which is one of its limitations. However, I demonstrate that the productivity of local service industries differs by competition or an agglomeration economy even though non-small firms have a possibility of a scale economy.

I also obtain the data on sales of retail trade, wholesale, output of manufacturing and the number of residents from the visible data of the Cabinet Office, Government of Japan; the data on employment by industry from the population census; the data on population by educational level and prefecture from the Employment Status Survey; and the data on elderly people from the Basic Resident Registration. Table 1 presents the detailed descriptive statistics (note: because of their sizes, all tables are found at the end of this paper).

Estimation Results

Factors That Affect the Productivity of the Service Sectors

Effect on Total Factor Productivity

Column 1 of Table 2 indicates that the high demand of retail. trade per resident increases the TFP of the retail trade industry. The large demand in retail trade might induce an intense agglomeration economy. In fact, column 9 indicates that the agglomeration economy increases the TFP of the retail trade industry. However, high income per resident does not directly increase the TFP. High density of establishments in retail trade increases the TFP of the retail trade industry, as indicated in columns 3 and 4. The thick markets of labor and inputs and the spillover effect from other establishments in the same industry increase the TFP. The square of density of establishments in retail trade is significant and negative, as indicated in column 4. Overcrowded establishments decrease the TFP.

The agglomeration of the manufacturing sector increases the TFP of the retail trade industry, as indicated in columns 5 and 6. High density of establishments in the manufacturing sector might increase the demand in retail trade, and high demand in retail trade increases its TFP. However, the spillover effect from manufacturing sectors might affect the retail trade industry. In fact, column 12 indicates that the magnitude of density of establishment in manufacturing is smaller when the demand of retail trade per resident (i.e. sales of retail trade per resident) is controlled. However, the coefficient of density of establishments in manufacturing is still significant. Therefore, manufacturing sectors improve the TFP of the retail trade industry by the spillover effect and increase in demand in retail trade.

Table 3 indicates similar results for the TFP of the wholesale industry. Remarkably, the intense competition indicates the significant and positive effect on TFP in the wholesale industry. The effect of. demand (i.e., sales per resident) in the wholesale industry is weaker than that in the retail trade industry (the coefficient is smaller, and columns 11 and 12 indicate insignificant coefficients). It might be the case that most customers in the wholesale industry are not households, but firms. Therefore, sales per resident do not affect TFP of the wholesale industry. The effect of density of establishments the in wholesale industry on its TFP is larger (e.g., the coefficient of the density of establishment in the retail trade industry in column 4 of Table 2 is 0.215, whereas, the coefficient of the density of establishment in the wholesale industry in column 4 of Table 3 is 0.605). Agglomeration economy also affects TFP of the wholesale industry, as indicated in columns 9 and 13. There are wholesale streets in Japan. It may be an expression of the effect from the agglomeration of the wholesale industry.

Table 4 indicates that in contrast to the above results, the density of establishments in both the narrowly defined service industries and manufacturing sector does not improve the TFP of the former. However, the agglomeration economy and competition improve its TFP. The high density of establishments does not increase the TFP, but the agglomeration of employment, regardless of workers working at the same place, improves the TFP. In Table 4, the ratio of elderly population is significant and negative while it has the opposite sign in Table 2 and 3. The narrowly defined service sector includes education, cinema, golf, amusement facilities, and temp service. Aging people may have a lower demand of these sectors; hence, it does not induce high competition and increase the TFP. I do not have data on the sales of narrowly defined service per resident.

Effect on Labor Productivity

Table 5 indicates that the square of the density of retail trade and the square of the density of manufacturing increases labor productivity. It implies that a certain density of manufacturing or retail trade improve labor productivity. The agglomeration economy has a negative impact on labor productivity. The TFP is increased by the agglomeration of establishments, and not the agglomeration of employment.

Effect on the Number of Workers

Table 6 indicates that the TFP has an insignificant impact on the change in the number of workers. However, labor productivity decreases the percentage change in the number of workers. It means that increase in labor productivity implies labor-saving progress. However, Morikawa (2016) argues that the flexibility of hiring workers corresponding to the demand in the service sector is one of the factors that improve its productivity. I do not determine whether new technology, improvement in production efficiency, management know-how, and other reasons increase the TFP and then the number of workers. Table 6 presents the relationship between productivity and the change in the number of workers.

Establishments That Enter or Exit the Market

Table 7 indicates that the high density of establishment in the retail trade and wholesale industries increase the percentage of entry to the market. However, the increase in the TFP of the retail trade industry decreases the percentage of entry to the market. Moreover, the increase in the density of retail trade and wholesale decrease the percentage of exit from the market. These results do not strongly support Darwinian selection. Kazekami (2016) also argues that the number of establishments that enter or exit the market is the most stagnant in the highest quartile of change in the TFP. As for the narrowly defined service industries, the increase in the TFP increases the percentage of entry to the market, but increase in the density of establishments in them increases the percentage of exit from the market. It might imply that Darwinian selection works for the narrowly defined service industries.

Municipalities Specializing in the Health and Nursing Sector and Agglomeration of Manufacturing Industries

Many rural municipalities have faced decreasing population and do not have a large manufacturing sector. Table 8 indicates that the health and nursing dummy (the health and nursing dummy equals one if a municipality specializes in this sector) is significant and negative for all service sectors (i.e., retail trade, wholesale, and the narrowly defined service sector). However, the cross-term. indicates that the agglomeration of establishments in the retail trade or wholesale industries increases the TFP even if the municipality specializes in the health and nursing sector, as indicated in columns 5 and 12. The cross-term between the health dummy and manufacturing indicates that agglomeration of establishments in the manufacturing sector does not improve the TFP of every service industry.

Figure 1 indicates the distribution of TFP of the retail trade industry in 2001 using municipal data. The TFP indicated by the red and blue lines, that is, the municipalities that do not specialize in the health and nursing sector, is higher than the TFP indicated by the green and orange lines, that is, the municipalities that specialize in the health and nursing sectors as there is a lack of manufacturing or core tradable sectors.

Comparing the blue distribution with the red distribution and the green distribution with the orange distribution, the municipalities that specialize in specific manufacturing industries have higher levels of TFP in the retail trade industry. It implies that the manufacturing or core tradable sectors are required to improve the productivity of the service sectors.

A few municipalities specializing in the specific manufacturing sector may have a bigger hospital and care houses. In that case, the health and nursing sector is also a specialized sector because people visit from other regions as well. These municipalities are indicated in the right tail of the orange line-distribution in Figure 1 (see next page)

Conclusion

I examined the effects of competition and an agglomeration economy on the productivity of service industries. First, an agglomeration economy improves the TFP of the retail trade industry. Second, competition improves the TFP of the wholesale industry and narrowly defined service industries. In Japan, some municipalities have few industries and the main industry is health and nursing. I find that these municipalities have lower productivity in the service sectors. However, agglomeration economies of retail trade, wholesale, and manufacturing industries increases the productivity even in these municipalities. These results suggest policy makers that encouraging the service sector only is not sufficient. They need to consider reconstructing all the regional industries.

FIGURE 1
Distribution of the TFP in Retail Trade in 2001

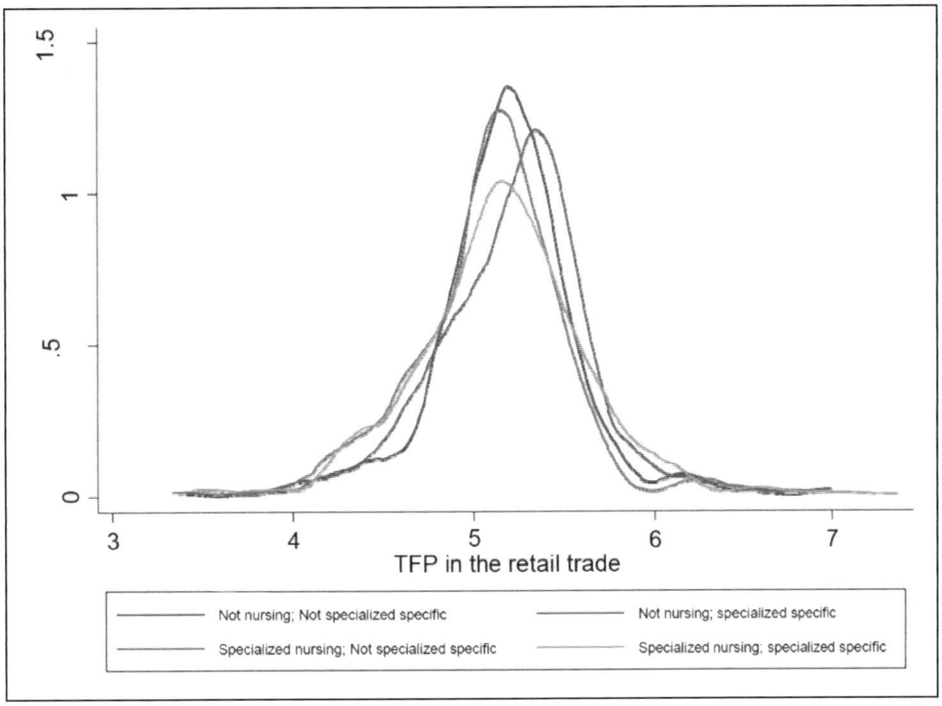

Not nursing = The municipalities do not specialize the health and nursing sectors.
Not specialized specific = The municipalities do not specialize the specific tradable industries.

Acknowledgments

The author thanks Daniel van Jaarsveld at University of British Columbia, Tracy Anderson at University of Pennsylvania and Tianlong You at Arizona State University, and participants in the Labor and Employment Relations Association the 69th Annual Meeting for their very helpful comments. The author is also grateful to Atsushi Seike, Atsushi Yashiro, and participants in Human Resource Management seminar at Keio University for useful comments. Moreover, the author is thankful for the financial support provided by the Japan Society for the Promotion of Science (grant number: 25380377).

Endnotes

[1] Labor data are available only for large classifications. Additionally, the estimation of TFP cannot be conducted for data using small classifications.

[2] This specialized index implies that, considering labor productivity is equal among municipalities, if a sector has a specialized index that is more than one, the sector trades surplus with other municipalities. I use the specialized index in each municipality published by the Ministry of Internal Affairs and Communications.

References

Ellison, Glen, and Edward L. Glaeser, 1997, "Geographic Concentration in U.S. Manufacturing Industries: a Dartboard Approach," *Journal of Political Economy*, Vol. 105, No. 5, pp. 889-927.

Foster, Lucia, John Haltiwanger, and C.J.Krizan, 2006, "Market Selection, Reallocation, and Restructuring in the U.S. Retail Trade Sector in the 1990s," *The Review of Economics and Statistics*, Vol. 88, No. 4., pp. 748-758.

Japan Revitalization Strategy, 2016, The cabinet in Japan.

Kazekami, Sachiko, 2016, "Do Service Sectors Need Another Core Sector to Improve Their Productivity?", presented at the 20th Eurasia Business and Economic Society Conference, the 14th Global Business and Economics Conference, and the 15th International Convention of the East Asian Economic Association.

Lucas, Robert E., 1988, "On the Mechanics of Economic Development," *Journal of Monetary Economics*, Vol. 22, pp. 3-41.

Moretti, Enrico, 2004, "Workers' Education, Spillovers, and Productivity: Evidence from Plant-Level Production Functions," *The American Economic Review*, Vol. 94, No. 3, pp. 656-690.

Morikawa, Masayuki, 2014, *Sabisu Sangyo No Seisansei Bunseki [Analysis of Productivity in Service Sectors]*, Nihon Hyoron Sya, Japan.

Morikawa, Masayuki, 2016, *Service-Oriented Nation [Sabisu Rikkoku Ron]*, Nihon Keizai shinbun syuppansya.

Rauch, James E., 1991, "Productivity Gains from Geographic Concentration of Human Capital: Evidence from the Cities," National Bureau of Economic Research (NBER) Working Paper No. 3905.

Syverson, Chad, 2004, "Market Structure and Productivity: A Concrete Example," *Journal of Political. Economy*, Vol. 112, No. 6, pp. 1181-1222.

Syverson, Chad, 2011, "What Determines Productivity?", *Journal of Economic Literature*, Vol. 49, No. 2, pp. 326-365.

Table 1 the descriptive statistics

Variable	Obs	Mean	Std. Dev.	Min	Max
TFP of retail trade	5932	5.251	0.490	0.974	7.810
TFP of wholesale	5433	6.361	0.749	1.817	11.652
TFP of the narrowly defined service industries	4775	4.915	0.752	1.491	11.161
Sales of retail trade per resident	5932	0.898	0.817	0.048	27.838
Sales of wholesale per resident	5433	2.232	33.498	0.000	1389.051
Income per resident	5932	1226.985	339.408	480.976	5508.200
Density of retail trade	5932	0.190	0.508	0.001	10.803
Density of wholesale	5433	0.055	0.349	0.000	10.836
Density of the narrowly defined service industries	4775	0.294	0.860	0.003	19.848
Density of manufacturing	5932	0.078	0.243	0.000	5.666
log establishment of retail trade	5932	5.912	1.161	2.773	11.399
log establishment of wholesale	4775	6.374	1.183	3.178	11.303
log establishment of the narrowly defined service industries	5433	4.069	1.483	-0.693	10.458
log establishment of manufacturing	5932	4.833	1.273	0.693	10.519
Agglomeration (retail trade and wholesale)	5932	0.000	0.000	0.000	0.000
Agglomeration (narrowly defined service industries)	4773	0.000	0.000	0.000	0.000
Competition (retail trade)	5932	0.271	0.280	0.013	1.000
Competition (wholesale)	5433	0.396	0.326	0.008	1.000
Competition (narrowly defined service industries)	4775	0.441	0.337	0.011	1.000
Labor productivity	5932	3.122	0.813	0.497	9.358
the change in employment of retial trade and wholesale	5917	-0.547	24.483	-108.942	92.789
Health and nursing dummy	4435	0.356	0.479	0.000	1.000
Ratio of persons with a college degree	5932	15.174	5.521	6.702	36.587
Ratio of elderly population	5932	22.183	6.594	5.717	47.373
percent of entry establishment in retail trade	2003	0.174	0.101	0.011	2.148
percent of entry establishment in wholesale	2003	0.206	0.122	0.000	1.667
percent of entry establishment in the narrowly defined service industri	2003	0.300	0.122	0.063	1.102
percent of exit establishment in retail trade	2003	0.190	0.119	0.000	0.467
percent of exit establishment in wholesale	2003	0.167	0.113	0.000	0.600
percent of exit establishment in the narrowly defined service industrie	2003	0.133	0.091	0.000	0.487

Table 2 the effect on TFP of the retail trade

	(1)	(2)	(3)	(4)	(5)	(6)	(7)	(8)	(9)	(10)	(11)	(12)	(13)	(14)
VARIABLES	Dependent variable=TFP of retail trade													
sales of retail trade per resident	0.100***										0.106**	0.0896**	0.100***	0.101***
	(0.0383)										(0.0439)	(0.0412)	(0.0382)	(0.0384)
Ratio of persons with a college degree	-0.0366***	-0.0398***	-0.0289***	-0.0237***	-0.0341***	-0.0315***	-0.0396***	-0.0403***	-0.0395***	-0.0398***	-0.0240***	-0.0322***	-0.0364***	-0.0367***
	(0.00505)	(0.00493)	(0.00565)	(0.00586)	(0.00539)	(0.00551)	(0.00509)	(0.00516)	(0.00498)	(0.00499)	(0.00583)	(0.00550)	(0.00505)	(0.00506)
Ratio of elderly population	0.00574	0.00583	0.00621*	0.00793**	0.00617*	0.00682*	0.00641	0.00558	0.00622*	0.00625*	0.00822**	0.00629*	0.00573	0.00576
	(0.00367)	(0.00378)	(0.00360)	(0.00366)	(0.00360)	(0.00361)	(0.00398)	(0.00404)	(0.00361)	(0.00362)	(0.00362)	(0.00372)	(0.00367)	(0.00368)
income per resident		-4.89e-05												
		(9.22e-05)												
density (retail trade)			0.0895***	0.215***							0.222***			
			(0.0168)	(0.0325)							(0.0311)			
square denstiy (retail trade)				-0.0136***							-0.0198***			
				(0.00293)							(0.00371)			
density (manufacturing)					0.187***	0.482***						0.325***		
					(0.0545)	(0.0866)						(0.110)		
square density (manufacturing)						-0.0560***						-0.0451***		
						(0.0129)						(0.0151)		
log establishment (retail trade)							0.00188							
							(0.0129)							
log establishment (manufacturing)								-0.00762						
								(0.0135)						
Agglomeration economy									9,204***				8,465***	
									(2,459)				(2,253)	
Competition										0.0168				0.0234
										(0.0860)				(0.0858)
Constant	5.261***	5.461***	5.235***	5.114***	5.305***	5.239***	5.374***	5.445***	5.386***	5.385***	5.009***	5.181***	5.257***	5.254***
	(0.0849)	(0.157)	(0.0911)	(0.0991)	(0.0889)	(0.0925)	(0.147)	(0.142)	(0.0841)	(0.0882)	(0.102)	(0.0955)	(0.0849)	(0.0890)
Observations	5,932	5,932	5,932	5,932	5,932	5,932	5,932	5,932	5,932	5,932	5,932	5,932	5,932	5,932
Number of city_code	1,558	1,558	1,558	1,558	1,558	1,558	1,558	1,558	1,558	1,558	1,558	1,558	1,558	1,558
Adjusted R-squared	0.363	0.358	0.361	0.363	0.360	0.360	0.358	0.358	0.358	0.358	0.367	0.364	0.363	0.363
city FE	Yes	Yes	Yes	Yes	Yes	Yes	Yes	Yes	Yes	Yes	Yes	Yes	Yes	Yes
year FE	Yes	Yes	Yes	Yes	Yes	Yes	Yes	Yes	Yes	Yes	Yes	Yes	Yes	Yes

Robust standard errors in parentheses
*** p<0.01, ** p<0.05, * p<0.1

density: density of establishment
log establishment: log of the number of establishments
Agglomeration economy: the modified Ellison-Glaeser measure of retail trade
Competition: the Herfindahl-Hirschman Index of retail trade

Table 3 the effect on TFP of the wholesale

VARIABLES	(1)	(2)	(3)	(4)	(5)	(6)	(7)	(8)	(9)	(10)	(11)	(12)	(13)	(14)
	Dependent variable=TFP of retail trade													
sales of wholesale per resident	0.000975***										-0.000140	0.000144	0.000966***	0.000909***
	(0.000252)										(0.000205)	(0.000272)	(0.000244)	(0.000234)
Ratio of persons with a college degree	0.0123	0.0114	0.0144*	0.0157**	0.0193**	0.0205**	0.0115	0.0125	0.0115	0.0122*	0.0158**	0.0204**	0.0125*	0.0132*
	(0.00747)	(0.00737)	(0.00761)	(0.00775)	(0.00816)	(0.00841)	(0.00754)	(0.00760)	(0.00741)	(0.00729)	(0.00775)	(0.00842)	(0.00748)	(0.00736)
Ratio of elderly population	-0.00123	-0.00185	-0.00137	-0.00112	-0.000939	-0.000639	-0.000364	0.000536	-0.000822	-0.000896	-0.00108	-0.000707	-0.00123	-0.00128
	(0.00382)	(0.00392)	(0.00380)	(0.00381)	(0.00379)	(0.00381)	(0.00415)	(0.00432)	(0.00380)	(0.00377)	(0.00382)	(0.00385)	(0.00382)	(0.00378)
income per resident		-0.000119												
		(0.000102)												
density (wholesale)			0.251***	0.605***							0.621***			
			(0.0683)	(0.128)							(0.138)			
square denstiy (wholesale)				-0.0249***							-0.0253***			
				(0.00721)							(0.00732)			
density (manufacturing)					0.250***	0.378***						0.369***		
					(0.0477)	(0.110)						(0.118)		
square density (manufacturing)						-0.0243*						-0.0236		
						(0.0143)						(0.0150)		
log establishment (wholesale)							0.00533							
							(0.0187)							
log establishment (manufacturing)								0.0156						
								(0.0191)						
Agglomeration economy									7,237***				7,074***	
									(2,524)				(2,510)	
Competition										0.453***				0.452***
										(0.0882)				(0.0882)
Constant	5.935***	6.115***	5.899***	5.861***	5.820***	5.789***	5.911***	5.827***	5.939***	5.755***	5.860***	5.791***	5.931***	5.747***
	(0.112)	(0.173)	(0.114)	(0.119)	(0.123)	(0.130)	(0.165)	(0.190)	(0.112)	(0.115)	(0.119)	(0.132)	(0.112)	(0.115)
Observations	5,433	5,433	5,433	5,433	5,433	5,433	5,433	5,433	5,433	5,433	5,433	5,433	5,433	5,433
Number of city_code	1,483	1,488	1,488	1,488	1,488	1,488	1,488	1,488	1,488	1,488	1,488	1,488	1,488	1,488
Adjusted R-squared	0.227	0.227	0.227	0.227	0.228	0.228	0.226	0.227	0.227	0.240	0.227	0.228	0.227	0.240
city FE	Yes	Yes	Yes	Yes	Yes	Yes	Yes	Yes	Yes	Yes	Yes	Yes	Yes	Yes
year FE	Yes	Yes	Yes	Yes	Yes	Yes	Yes	Yes	Yes	Yes	Yes	Yes	Yes	Yes

Robust standard errors in parentheses

*** p<0.01, ** p<0.05, * p<0.1

density: density of establishment

log establishment: log of the number of establishments

Agglomeration economy: the modified Ellison-Glaeser measure of wholeslae

Competition: the Herfindahl-Hirschman Index of wholeslae

Table 4 the effect on TFP of the narrowly defined service

	(1)	(2)	(3)	(4)	(5)	(6)	(7)	(8)	(9)
VARIABLES	Dependent variable=TFP of the narrowly service								
Ratio of persons with a college degree	-0.00747	-0.00683	-0.00723	-0.00635	-0.00636	-0.00713	-0.00413	-0.00826	-0.0105
	(0.00973)	(0.0104)	(0.0107)	(0.0107)	(0.0109)	(0.00994)	(0.0102)	(0.00969)	(0.00969)
Ratio of elderly population	-0.0145**	-0.0108*	-0.0108*	-0.0106*	-0.0106*	-0.00700	-0.00577	-0.0105*	-0.0112**
	(0.00597)	(0.00574)	(0.00574)	(0.00569)	(0.00571)	(0.00623)	(0.00626)	(0.00568)	(0.00567)
income per resident	-0.000377**								
	(0.000147)								
density (narrowly defiend service)		-0.0300	-0.0177						
		(0.0231)	(0.0629)						
square denstiy (narrowly defined service)			-0.000610						
			(0.00208)						
density (manufacturing)				0.0724	0.0709				
				(0.0581)	(0.126)				
square density (manufacturing)					0.000278				
					(0.0190)				
log establishment (narrowly defined service)						0.0370			
						(0.0255)			
log establishment (manufacturing)							0.0521**		
							(0.0250)		
Agglomeration economy								13,067***	
								(4,899)	
Competition									0.173**
									(0.0703)
Constant	5.586***	5.013***	5.015***	4.988***	4.988***	4.722***	4.623***	5.019***	4.949***
	(0.241)	(0.138)	(0.140)	(0.148)	(0.157)	(0.275)	(0.260)	(0.136)	(0.140)
Observations	4,775	4,775	4,775	4,775	4,775	4,775	4,775	4,773	4,775
Number of city_code	1,398	1,398	1,398	1,398	1,398	1,398	1,398	1,398	1,398
Adjusted R-squared	0.111	0.109	0.109	0.109	0.109	0.110	0.110	0.109	0.112
city FE	Yes	Yes	Yes	Yes	Yes	Yes	Yes	Yes	Yes
year FE	Yes	Yes	Yes	Yes	Yes	Yes	Yes	Yes	Yes

Robust standard errors in parentheses
*** p<0.01, ** p<0.05, * p<0.1

density: density of establishment
log establishment: log of the number of establishments
Agglomeration economy: the modified Ellison-Glaeser measure of narrowly defined service
Competition: the Herfindahl-Hirschman Index of narrowly defined service

Table 5 The effect on the labor productivity

VARIABLES	(1)	(2)	(3)	(4)	(5)	(6)	(7)	(8)	(9)	(10)	(11)	(12)	(13)	(14)	(15)
	Dependent variable=log of labor productivity														
sales of retail trade per resident	0.147											0.266***	0.302***	0.148	0.146
	(0.0992)											(0.0880)	(0.0696)	(0.0996)	(0.0990)
Ratio of persons with a college degree	0.0925***	0.0885***	0.0393***	0.00746	0.0551***	0.0364***	0.0433***	0.0341***	0.0875***	0.0884***	0.0909***	0.00662	0.0339***	0.0920***	0.0928***
	(0.00934)	(0.00895)	(0.0110)	(0.0108)	(0.00990)	(0.00980)	(0.00593)	(0.00582)	(0.00945)	(0.00945)	(0.0100)	(0.0109)	(0.00979)	(0.00936)	(0.00936)
Ratio of elderly population	0.0862***	0.0921***	0.0871***	0.0766***	0.0873***	0.0827***	0.0168***	0.0206***	0.0870***	0.0868***	0.0892***	0.0773***	0.0809***	0.0862***	0.0861***
	(0.00577)	(0.00577)	(0.00553)	(0.00543)	(0.00551)	(0.00546)	(0.00411)	(0.00396)	(0.00570)	(0.00569)	(0.00602)	(0.00540)	(0.00540)	(0.00577)	(0.00576)
income per resident		0.000633***													
		(0.000104)													
density (retail trade)			-0.404***	-1.165***								-1.145***			
			(0.0731)	(0.109)								(0.124)			
square denstiy (retail trade)				0.0825***								0.0669***			
				(0.0134)								(0.0140)			
density (manufacturing)					-1.102***	-3.188***							-3.716***		
					(0.235)	(0.364)							(0.410)		
square density (manufacturing)						0.397***							0.433***		
						(0.0676)							(0.0661)		
log establishment (retail trade)							-0.759***								
							(0.0205)								
log establishment (manufacturing)								-0.778***							
								(0.0183)							
Agglomeration economy									-16,383***					-17,475***	
									(4,184)					(4,546)	
Competition										-0.122*	-0.0162				-0.113
										(0.0741)	(0.0571)				(0.0726)
Constant	0.478***	-0.249	1.370***	2.102***	1.172***	1.639***	7.061***	6.259***	0.675***	0.702***	0.661***	1.837***	1.447***	0.485***	0.511***
	(0.180)	(0.224)	(0.186)	(0.185)	(0.167)	(0.172)	(0.202)	(0.172)	(0.152)	(0.153)	(0.159)	(0.195)	(0.182)	(0.180)	(0.181)
Observations	5,932	5,932	5,932	5,932	5,932	5,932	5,932	5,932	5,932	5,932	5,439	5,932	5,932	5,932	5,932
Number of city_code	1,558	1,558	1,558	1,558	1,558	1,558	1,558	1,558	1,558	1,558	1,492	1,558	1,558	1,558	1,558
Adjusted R-squared	0.287	0.287	0.315	0.347	0.307	0.336	0.608	0.618	0.281	0.281	0.289	0.363	0.360	0.287	0.287
city FE	Yes	Yes	Yes	Yes	Yes	Yes	Yes	Yes	Yes	Yes	Yes	Yes	Yes	Yes	Yes
year FE	Yes	Yes	Yes	Yes	Yes	Yes	Yes	Yes	Yes	Yes	Yes	Yes	Yes	Yes	Yes

Robust standard errors in parentheses

*** p<0.01, ** p<0.05, * p<0.1

density: density of establishment

log establishment: log of the number of establishments

Agglomeration economy: the modified Ellison-Glaeser measure of retail trade

Competition: the Herfindahl-Hirschman Index of retail trade in (10) and (15), and that index of wholesale in (11)

Table 6 the effect of productivity on the percent change in the number of workers

	(1)	(2)	(3)	(4)	(5)	(6)	(7)	(8)
VARIABLES	Dependent variable=the percent changein the number of workers							
TFP in the retail trade	-0.402	-0.428						
	(1.020)	(1.030)						
TFP in the whwolesale			0.254	0.258				
			(0.821)	(0.823)				
average TFP					0.291	0.306		
					(1.381)	(1.394)		
Labor productivity							-35.51***	-36.25***
							(0.895)	(0.948)
sales of retail trade per resident		0.236		-0.378		-0.387		10.19**
		(1.225)		(1.289)		(1.299)		(4.034)
sales of wholesale per resident		0.0102		0.0204		0.0206		-0.257***
		(0.0400)		(0.0418)		(0.0419)		(0.0646)
Ratio of persons with a college degree	-1.972***	-1.955***	-2.178***	-2.169***	-2.171***	-2.162***	1.177***	1.303***
	(0.338)	(0.342)	(0.360)	(0.367)	(0.361)	(0.367)	(0.301)	(0.294)
Ratio of elderly population	-2.420***	-2.425***	-2.447***	-2.453***	-2.448***	-2.454***	0.670***	0.783***
	(0.281)	(0.283)	(0.311)	(0.314)	(0.311)	(0.314)	(0.241)	(0.249)
Constant	75.19***	74.95***	74.53***	74.84***	74.38***	74.64***	96.60***	85.90***
	(8.905)	(8.959)	(8.806)	(8.948)	(10.92)	(10.96)	(5.814)	(7.005)
Observations	5,917	5,917	5,262	5,262	5,262	5,262	5,917	5,917
Number of city_code	1,558	1,558	1,444	1,444	1,444	1,444	1,558	1,558
Adjusted R-squared	0.056	0.056	0.056	0.055	0.056	0.055	0.371	0.377
city FE	Yes	Yes	Yes	Yes	Yes	Yes	Yes	Yes
year FE	Yes	Yes	Yes	Yes	Yes	Yes	Yes	Yes

Robust standard errors in parentheses
*** p<0.01, ** p<0.05, * p<0.1

Table 7 the establishments enter or exit the market

	(1)	(2)	(3)	(4)	(5)	(6)	(7)	(8)	(9)	(10)	(11)	(12)	(13)	(14)	(15)
VARIABLES								The percent of entry the market							
The change in TFP of retail trade	-0.0172**	-0.0162*	-0.0165**												
	(0.00832)	(0.00827)	(0.00813)												
density (retail trade)		0.0203***	0.00705		0.00740	0.0206***									
		(0.00377)	(0.00352)		(0.00855)	(0.00382)									
sales of retail trade per resident			0.0734	0.0781	0.0733										
			(0.0557)	(0.0513)	(0.0559)										
The change in TFP of wholesale							0.000887	0.000989	0.000925						
							(0.00986)	(0.00986)	(0.00986)						
density (wholesale)								0.147***	0.183***		0.183***	0.147***			
								(0.0362)	(0.0505)		(0.0505)	(0.0362)			
sales of wholesale per resident									-0.000473***	3.24e-05	-0.000474***				
									(0.000151)	(0.000139)	(0.000151)				
The change in TFP of the service*													0.00842*	0.00838*	
													(0.00469)	(0.00469)	
density (service)														0.0113	0.0115
														(0.0137)	(0.0139)
Adjusted R-squared	0.550	0.553	0.565	0.565	0.565	0.552	0.059	0.063	0.064	0.059	0.064	0.064	0.652	0.652	0.650
VARIABLES								the percent of exit the market							
The change in TFP of retail trade	0.00869	0.00546	0.00542												
	(0.00830)	(0.00793)	(0.00795)												
density (retail trade)		-0.0428***	-0.0444***		-0.0445***	-0.0429***									
		(0.0101)	(0.0102)		(0.0102)	(0.0101)									
sales of retail trade per resident			0.00920	-0.0201***	0.00923										
			(0.00826)	(0.00653)	(0.00827)										
The change in TFP of wholesale							0.00364	0.00353	0.00354						
							(0.00701)	(0.00701)	(0.00701)						
density (wholesale)								-0.103*	-0.106		-0.106	-0.103*			
								(0.0614)	(0.0675)		(0.0674)	(0.0613)			
sales of wholesale per resident									4.64e-05	-0.000250***	4.42e-05				
									(0.000158)	(4.65e-05)	(0.000157)				
The change in TFP of the service													-0.000435	-0.000510	
													(0.00201)	(0.00199)	
density (service)														0.0417***	0.0417***
														(0.00741)	(0.00740)
Observations	2,003	2,001	2,001	2,003	2,001	2,001	2,003	2,001	2,001	2,003	2,001	2,001	2,003	2,001	2,001
Number of city_code	1,102	1,102	1,102	1,102	1,102	1,102	1,102	1,102	1,102	1,102	1,102	1,102	1,102	1,102	1,102
Adjusted R-squared	0.771	0.789	0.790	0.772	0.789	0.789	0.635	0.637	0.637	0.635	0.637	0.637	0.768	0.781	0.781
city FE	Yes	Yes	Yes	Yes	Yes	Yes	Yes	Yes	Yes	Yes	Yes	Yes	Yes	Yes	Yes
year FE	Yes	Yes	Yes	Yes	Yes	Yes	Yes	Yes	Yes	Yes	Yes	Yes	Yes	Yes	Yes

Robust standard errors in parentheses

*** $p<0.01$, ** $p<0.05$, * $p<0.1$

density: density of establishment

*service: the narrowly defined service industries

Table 8 Health and nursing sector and agglomeration of manufacturing

	(1)	(2)	(3)	(4)	(5)	(6)	(7)	(8)	(9)	(10)	(11)	(12)	(13)	(14)	(15)	(16)	(17)	(18)	(19)	(20)
VARIABLES	Dependent variable=TFP of retail trade							Dependent variable=TFP of wholesale							Dependent variable=TFP of narrow service					
Health and nursing dummy	-0.0738***	-0.0640***	-0.0673***	-0.0590**	-0.0821***	-0.0692***	-0.0826***	-0.106***	-0.103***	-0.0941**	-0.0972**	-0.138***	-0.101***	-0.101**	-0.105**	-0.0942**	-0.0878**	-0.112**	-0.0999**	-0.107**
	(0.0230)	(0.0232)	(0.0230)	(0.0230)	(0.0268)	(0.0230)	(0.0262)	(0.0379)	(0.0379)	(0.0378)	(0.0381)	(0.0460)	(0.0379)	(0.0434)	(0.0425)	(0.0425)	(0.0428)	(0.0560)	(0.0425)	(0.0507)
sales of retail trade per resident		0.0594***																		
		(0.0174)																		
sales of wholesale per resident									0.00107***											
									(0.000345)											
Ratio of persons with a college degree	0.00461**	0.00351*	-0.00281	-0.000750	-0.000834	0.000651	0.000576	0.0103***	0.00935***	-0.000149	0.00694*	0.00615*	0.00669*	0.00664*	-0.00256	-0.0109***	-0.0105***	-0.0108***	-0.00649*	-0.00672*
	(0.00198)	(0.00200)	(0.00253)	(0.00206)	(0.00207)	(0.00210)	(0.00211)	(0.00337)	(0.00337)	(0.00420)	(0.00368)	(0.00371)	(0.00369)	(0.00371)	(0.00341)	(0.00406)	(0.00387)	(0.00388)	(0.00351)	(0.00351)
Ratio of elderly population	-0.00373*	-0.00347*	0.000497	-0.00223	-0.00173	-0.00271	-0.00245	-0.0196***	-0.0199***	-0.0141***	-0.0198***	-0.0185***	-0.0188***	-0.0188***	0.00981***	0.0144***	0.0100***	0.0107***	0.0107***	0.0109***
	(0.00206)	(0.00205)	(0.00231)	(0.00208)	(0.00211)	(0.00208)	(0.00211)	(0.00311)	(0.00311)	(0.00322)	(0.00311)	(0.00318)	(0.00312)	(0.00314)	(0.00358)	(0.00375)	(0.00356)	(0.00369)	(0.00362)	(0.00369)
income per resident			0.000217***							0.000297***						0.000241***				
			(4.87e-05)							(6.94e-05)						(7.39e-05)				
density (retail trade)				0.180***	0.169***															
				(0.0235)	(0.0238)															
square denstiy (retail trade)				-0.0101***	-0.00895***															
				(0.00234)	(0.00227)															
Health × density (retail trade)					0.220**															
					(0.0991)															
Health × square denstiy (retail trade)					-0.0786**															
					(0.0373)															
density (manufacturing)						0.305***	0.282***						0.249***	0.256***					0.272***	0.283***
						(0.0646)	(0.0691)						(0.0743)	(0.0816)					(0.0888)	(0.0963)
square denstiy (manufacturing)						-0.0383***	-0.0339***						-0.0218*	-0.0229*					-0.0291*	-0.0302*
						(0.0116)	(0.0117)						(0.0125)	(0.0133)					(0.0169)	(0.0180)
Health × density (manufacturing)							0.265							-0.000441						0.183
							(0.227)							(0.256)						(0.301)
Health × square denstiy (manufacturing)							-0.126							-0.0253						-0.179
							(0.114)							(0.124)						(0.153)
density (wholesale)											0.236***	0.224***								
											(0.0761)	(0.0742)								
square denstiy (wholesale)											-0.0120*	-0.0106								
											(0.00717)	(0.00691)								
Health × density (wholesale)												2.617***								
												(0.903)								
Health × square denstiy (wholesale)												-5.092**								
												(2.207)								
density (service)																	0.138***	0.136***		
																	(0.0248)	(0.0250)		
square denstiy (service)																	-0.00416***	-0.00403***		
																	(0.00117)	(0.00119)		
Health × density (service)																		0.219		
																		(0.180)		
Health × square denstiy (service)																		-0.116		
																		(0.0820)		
Constant	5.052***	4.998***	4.786***	5.045***	5.040***	5.056***	5.054***	6.269***	6.282***	5.910***	6.299***	6.289***	6.276***	6.276***	4.642***	4.342***	4.705***	4.698***	4.648***	4.647***
	(0.0518)	(0.0530)	(0.0819)	(0.0516)	(0.0517)	(0.0518)	(0.0518)	(0.0805)	(0.0805)	(0.114)	(0.0818)	(0.0817)	(0.0807)	(0.0807)	(0.0860)	(0.127)	(0.0861)	(0.0865)	(0.0854)	(0.0855)
Observations	4,435	4,435	4,435	4,435	4,435	4,435	4,435	3,966	3,966	3,966	3,966	3,966	3,966	3,966	3,437	3,437	3,437	3,437	3,437	3,437
Number of city_code	1,536	1,536	1,536	1,536	1,536	1,536	1,536	1,427	1,427	1,427	1,427	1,427	1,427	1,427	1,316	1,316	1,316	1,316	1,316	1,316
year FE	Yes	Yes	Yes	Yes	Yes	Yes	Yes	Yes	Yes	Yes	Yes	Yes	Yes	Yes	Yes	Yes	Yes	Yes	Yes	Yes

Robust standard errors in parentheses

*** p<0.01, ** p<0.05, * p<0.1

density: density of establishment

others: the other service sectors

service:the narrowly defined service industries

VIII. 2017 Lifetime Achievement Award

Comments

JOHN F. BURTON, JR.
Professor Emeritus
Rutgers University and Cornell University

Albert Rees, then a distinguished labor economist at the university of Chicago, provided this analysis of the role of unions (1962, 194-95):

> If the union is viewed solely in terms of its effect on the economy, it must in my opinion be considered an obstacle to the optimum performance of our economic system. It alters the wage structure in a way that impedes the growth of employment in sectors of the economy where productivity and income are naturally high and that leaves too much labor in low income sectors of the economy like southern agriculture and the least skilled service trades. It benefits most those workers who would in any case be relatively well off, and while some of this gain may be at the expense of the owners of capital, most of it must be at the expense of consumers and the lower-paid workers. Unions interfere blatantly with the use of the most productive techniques in some industries, and this effect is probably not offset by the stimulus to higher productivity furnished by some other unions.

Many of my fellow economists would stop at this point and conclude that unions are harmful and that their power should be curbed. I do not agree that one can judge the value of a complex institution from so narrow a point of view. Other aspects of unions must also be considered. The protection against the abuse of managerial authority given by the seniority systems and grievance procedures seems to me to be a union accomplishment of the greatest importance. So too is the organized representation in public affairs given the worker by the political activities of unions. If, as most of us believe, American should continue to have political democracy and a free enterprise economy it is essential that the great mass of manual workers be committed to the preservation of this system and that they should not, as in many other democracies, constantly be attempting to replace it with something radically different. Yet such a commitment cannot exist if workers feel that their rights are not respected and they do not get a fair share of the rewards of the system. By giving workers protection against arbitrary treatment by employers, by acting as their representative in politics, and by reinforcing their hope of continuous future gain, unions have helped assure that the basic values of our society are widely diffused and that our disagreements on political and economic issues take place within a broad framework of agreement. If the job rights won for workers by unions are not conceded by the rest of society simply because they are just, they should be conceded because they help to protect the minimum consensus that keeps our society stable. In my judgment, the economic losses imposed by unions are not too high a price to pay for their successful performance of this role.

I would restate the analysis by Rees and identify three contributions of unions. First, unions help their members by improving wages and working conditions and by protecting workers from arbitrary treatment by management. Second, unions also help all workers by establishing reasonable standards for wages and compensation through collective bargaining that unorganized firms are to some extent required to emulate. And third, as aptly described by Rees, by helping workers share in the increasing wealth of the economy, unions contribute to the stability of our society. In essence, unions should be endorsed and supported by conservatives because unions support rather than undermine the features essential to the long-term survival of our country.

Author's address: Rutgers School of Management and Labor Relations, 94 Rockefeller Road, Piscataway Township, NJ 08854

I used to assign the analysis of unions by Rees to my students in the MBA program at the University of Chicago. Alas, I don't think that many were persuaded that conservatives or members of the business community should endorse a robust union movement. But this goal is so important for our country that I want to try again to persuade some businessmen that resistance to viable unions is a myopic policy.

The Labor and Employment Relations Association has a long history of promoting constructive labor-management relations. I propose that LERA establish an annual award that recognizes a member of the business community who has made a contribution to the continuation of a viable union movement in the U.S.

The award should be named the Myron Taylor Award.[1] Myron Taylor was Chairman of U.S. Steel who in 1937 signed a collective e bargaining agreement with the SWOC (the Steel Workers' Organizing Committee), which John H. Lewis had been instrumental in founding.

I do not want to suggest that Myron Taylor specifically endorsed the proposition that unions are essential to the long-run survival of the country. His motivation was more practical: "that unions were an unavoidable reality in the future of steel production, and that continued resistance would only delay the inevitable while exposing the company to considerable political and economic risks" (White 2016, 101). But regardless of Taylor's motive, to his credit, Taylor made his decision to recognize the SWOC when the leaders of the rest of the steel industry persisted in their vigorous resistance to unionization. Indeed, the organizing campaign in the rest of the steel industry—known as Little Steel—that followed the recognition of the UWOC by U.S. Steel resulted in a series of violent confrontations that have been termed "The Last Great Strike" in American labor history (White 2016).[2] Moreover, in some ways, Myron Taylor was even ahead of workers in his recognition of the reality of unionization. He signed the collective bargaining agreement with the SWOC even though "the union had actually signed up only 7 percent of U.S. Steel's workers when Taylor's negotiations with Lewis unfolded" (White 2016, 101).

And so let us reserve a chair at the head table at next year's LERA awards luncheon for the recipient of the Myron Taylor Award. Needless to say, the award may not be given every year. But let us look forward to years when we will need a longer table to accommodate the multiple recipients of the award.

Endnotes

[1] Myron Charles Taylor (1874-1959) was an industrialist, a diplomat, and a philanthropist. He was a 1894 graduate of the Cornell Law School, and he contributed $1.5 million in 1928 to Cornell University to construct a new building for the Cornell Law School and Law Library. In 1949, he made an additional gift of $1.5 million to Cornell to build Anabel Taylor Hall, which was named in his wife's honor and which serves as an interdenominational religious center (http://bit.ly/2mSEJcW2017).

[2] The best-known confrontation in the effort to organize Little Steel is known as the Memorial Day Massacre, which took place at Republic Steel's South Chicago Works on May 30, 1937. The police fired about 200 gunshots at the demonstrators that resulted in mortal wounds to ten men and nonlethal wounds to some thirty men and women (White 106, 2-3).

There were several other violent confrontations involving Republic Steel and demonstrators, including what has been called "The Riot at Massillon" (White 2016, 211), which took place at the Massillon, Ohio plant on July 7, 1937 and which resulted in two deaths and injuries to about a dozen persons, mostly from gunshots wounds (White 2016, 216-17). I have a particular interest in this incident because I grew up in Massillon and worked two summers at Republic Steel in the 1950s when I was in college. Alas, the Massillon Steel Works has long since been closed.

A more significant reason for my continuing interest is that my sister, Sue Burton, wrote a poem about the 1937 demonstrations in Massillon. The first two stanzas of *Little Steel* commemorates the lives of the two workers killed by the police (Burton 2016):

> Let us praise Fulgencia Calzada, shot
> In the back of the head. Let us interrogate the bullet.
> *Oh, but the strikers threw a rock.*

And let us praise Nick Vathias—or Vadios
Or Vadlas—gunned down at the door to the strikers' kitchen
Excusable homicide.

References

Burton, Sue D. 2016. "Little Steel." Mudlark 60. http://bit.ly/2Gaeu9B

Rees, Albert. 1962. *The Economics of Trade Unions.* Chicago, IL: University of Chicago Press.

White, Ahmed. 2016. *The Last Great Strike: Little Steel, the CIO, and the Struggle for Labor Rights in New Deal America.* Oakland, CA: University of California Press

Remarks

SIR PETER CARR
English National Health Service (retired)

It was an immense privilege to receive this Lifetime Achievement award from LERA. I am really sorry that I cannot be there in person to thank you. My friendship and work with LERA and American colleagues has been an enormous pleasure over the years.

I started my working life when I left school at fourteen and took an apprenticeship as a carpenter and joiner, working in the building industry. This early working experience shaped my future career. It has been a lifetime of studying, teaching, organising, regulating, mediating and managing, in support of the wider public interest.

My early industrial relations work was in key British government institutions, as well as extended work in the United States and Sweden, as well as evidence gathering in many other countries.

From the 1950s to the late 1970s, Britain suffered from a chaotic bargaining structure and deep antipathy between unions and employers. With hindsight, we might now judge that it was an impossible task to successfully engage in a consensual reform of the system as it was. But for a group of us who left the teaching of industrial relations to engage in the reform programme of the late 1960s, the opportunity to be directly involved was a dream come true.

I came into industrial relations practice from the Oxford School, under the tuition of Alan Flanders and Hugh Clegg. Back then, we nurtured the concept of rational structures, encouraging a more articulate and effective contribution by trade unions, and an adherence to mutually agreed procedures. Central to this philosophy was the idea that collective bargaining was the best means of determining relationships between management and labour.

The main institution that came out of this Oxford School thinking was the Commission on Industrial Relations (CIR). The CIR took me on as a director and we set out to produce effective procedures and conflict resolution methods in companies and industries with the most disruptive histories. I had previously been employed part-time at the National Board for Prices and Incomes (NBPI), which attempted to reform pay determination and aimed to spread the practice of productivity bargaining to the benefit the employer, the majority of employees and to the consumer. The pay bargaining experience gained at the NBPI provided skills that we deployed in other national reform initiatives at the CIR. It was a rational, researched and liberal approach to industrial relations. Although it may have had only a partial success, I believe that it could provide vital lessons worth considering today, as we move into a new era of disruptive technologies that are again set to radically change the world of work and working relations.

The introduction of a legal framework in Britain, in many ways mirroring the American system, dominated all our agendas in the 1960s and early 70s. Back then, national strikes could erupt on matters of fine principle. Yet we established a significant methodology to deal with the problems. For example, with a team of researchers I spent over a year in the national newspaper industry for the Royal Commission on the Press, to form a new agenda for industrial relations in that industry. The thorough nature of that study continues to receive praise to this day. The case study methodology we used in union recognition cases was reviewed in a recent study. In the end, a changed political environment enabled the newspaper employers to sort out their industry in their own way, and in their favour.

My keen interest in Worker Participation led me to travel to six European countries to study their various organisational forms. We took some elements of that learning through the UK Parliament, including

Sir Peter Carr was not able to be physically present at the June 2017 conference in Anaheim to accept the Lifetime Achievement Award, but he sent us his acceptance remarks, which detail his involvement with colleagues in the United States and the Washington LERA chapter. He subsequently passed away in October 2017 at the age of 87. His obituary can be found at https://lera.memberclicks.net/sir-peter-carr-obituary

the introduction of standards applying to the provision of information for union representatives. Of course, such provisions have no legal effect these days.

In the 1970s, I was repeatedly drawn to America as a country from which to learn from the experience of free collective bargaining within a strong legal framework. In 1976, the American Ambassador in Britain, Ann Armstrong, provided a generous opportunity to undertake a study tour in America to broaden my experience of the way the American systems operate.

I learned a lot in that period from the experience of American academics, trade unions and employers. During that tour, I had a memorable meeting with George Meany and developed a relationship which we retained for many years. We also held several interesting debates in Oxford with Walter and Gladys Gershenfeld. I even attempted, but failed, to recruit Tia Denenberg to the UK during Richard Denenberg's study period in Britain. I had more success later with Eileen Hoffman's recruitment into the UK Conciliation Service (ACAS) who came on an exchange scheme that I established with the FNCS.

My appointment as Counsellor for Labour with the British Embassy in Washington came in 1978. By the time I arrived in Washington, I already had many American friends and colleagues. The job took me to every state and brought me into contact with many of the great figures in American labour history. I met A. Philip Randolph, whose organising skills had established the rights of sleeping-car porters. I was also able to meet Bayard Rustin who told me about his role in organising the Great March on Washington. I also learned a great deal from Victor Reuther of his and Walter Reuther's organisation of the autoworkers.

A major task I undertook during my Washington British Embassy tenure involved me going into American Universities, from which I produce a series of papers on the functioning of collective bargaining in the US. Those papers, in turn, re-shaped the British government's thinking of its own industrial relations system in the early 1980s. Altogether, I think that I must have produced over two hundred papers on aspects of the American Labour market. Looking back through those papers recently, it is clear how much has changed since then. They were written before the profound effect of new technology on the labour market. However, some pivotal assumptions continue to be relevant. I am convinced that the relationship of employer and employee remains a pluralistic process, and that to remain healthy this requires effective employee representation, whatever the technology. We should now be experimenting with new ways for employee views to be articulated in the modern workplace.

In 1982, during my time in Washington, the American Labor Department persuaded me to participate in a collective bargaining conference they were sponsoring in Rio de Janeiro. This was before the fully democratic elections in Brazil. It had been arranged for me to talk to a shop-stewards group at the Volkswagen car plant, to which the Brazilian security police raised objections. I came back from Brazil reinforced in the conviction that trade unions are an essential component for a truly democratic society. A senior shop steward called Luiz Inácio Lula da Silva was a participant on that course. He later went on to become the democratically elected President of the Brazilian Republic.

On completion of the Washington diplomatic post, I returned to Britain to head a team to deal with the impact of the 1980s recession and its consequent devastation of the industrial infrastructure, supporting the regeneration of the North-East economy. My experience in America provided many lessons for establishing effective community responses to such rapid industrial change.

I took retirement from government service around twenty years ago but was soon encouraged back to work again, this time with the British National Health Service (the NHS), and in turn headed up five major health organisations. I am proud to say that at one point the North-East health system received the accolade of being the best operating part of the National Health Service. In my work for these health bodies, the discipline and constraints acquired from my industrial relations experience have been essential.

Finally, I would like to thank LERA for its friendship over the years, including Janice Bellace who brought this wonderful award over to the UK for me. In particular, I would like to extend a personal thanks to Greg Bamber in Australia who has been a constant sounding board for many years since we first worked together at the Commission on Industrial Relations.

LERA stands in a good position to help define the boundaries and the processes that will be needed for the challenging new economy ahead. As a forum for exchanging experience and ideas, it has a unique and vital role to play. Long may you continue with your important work!

VIII. Dynamics of Low-Wage Labor Markets: Implications for Minimum Wage Policy

When Hours Decline: Tides of Change in Low-Wage Labor Markets

WENDY RAYACK
Wesleyan University

Using a sample from the PSID for 1997–2013, we document features that characterize flows of workers through low-wage markets over time. We show that half of the population moves through these markets at some point in their lives. Longitudinal analysis reveals a constant churning of the labor force and multiple strategies for adjusting to shocks, traits often masked by cross-section snapshots. In such an environment, higher wages could promote attachment, bring lower turnover, and yield gains in productivity, prospects that suggest new ways to think about predicted consequences of minimum-wage hikes.

Dramatic changes in the U.S. labor market constitute a growing "national crisis," according to many accounts. Those reports point to men of prime working age who have no steady connection to paid work and men who have dropped out of the labor force entirely[1] (Eberstadt 2016). Parallel stories report both women and men unable to sustain families on minimum-wage earnings. Both developments suggest an urgent need for attention to these markets. While some contend that hiking the minimum wage would help, others worry that a higher minimum wage would only push jobs further out of reach for those least prepared for the modern world of work[2] (Belman and Wolfson 2014; Neumark and Wascher 2008). We argue that these debates would benefit from a closer look at how low-wage labor markets operate when responding to shocks. Are jobs simply wiped out, or are hours and work conditions altered to reduce layoffs? Do shocks fall on otherwise stable low-wage markets or are such markets typified by volatility and a constantly changing workforce? To what extent are workers dropping out of the workforce completely as opposed to cycling in and out for a variety of reasons?

The growing literature on puzzles confronting minimum-wage research gives these questions heightened importance. Why, for example, do minimum-wage studies frequently find small and insignificant impacts on jobs? Could the volatility faced by low-wage sectors spur "multiple paths of adjustments" and explain the small to insignificant jobs effects found in some minimum-wage studies3 (Hirsch, Kaufman, and Zelenska 2015)? Would the most vulnerable people still lose an already tenuous attachment to steady work, or would higher wages transform secondary jobs, with negative traits, into jobs more like those in the primary sector, with increased worker attachment and less costly turnover?

In addressing these issues, we argue that snapshots of the labor market based on short-term cross-section data can hide central features of markets. In particular, they risk missing the strategies for adjusting to shocks and understating the flows of workers through these markets over time. To capture such changes, we use data from the Panel Study of Income Dynamics (PSID) for the period 1997 to 2013 and document the alterations in labor market status as work lives and family lives unfold.

In recent years, the PSID has collected information every other year rather than annually. Yet that limitation is outweighed by the ability to follow the same people from survey to survey and to include years before, during, and after the Great Recession. Exploiting the panel nature of the data, we explore what can be learned by following people as they navigate their way in and out of low-wage jobs over a substantial period

Author's address: Wesleyan University, Public Affairs Center, 327, 45 Wyllys Avenue, Middletown, CT 06459 of time. Our weighted sample includes observations on about 7,000 individuals per year, ages 19 to 64 (Appendix Table A1 and Appendix Table A2). We take advantage of the PSID's particularly high level of detail on the labor market activities of wives, husbands, and single women who head households.

Section I explores the distinction between permanent and temporary low-wage status. Section II compares transition patterns among labor market categories for workers starting from low- and higher-wage jobs. Section III links the observed patterns to increased labor market slack during the Great Recession and tracks long-run sequences of labor force exits and re-entry. Section IV makes the statistical links explicit. Using regression analysis, we estimate the magnitude and statistical significance of adjustments that occur in both high- and low-wage markets when shocks cause hours to decline.

Permanent Versus Temporary Low-Wage Status

It is common to talk about low-wage workers as if such workers come branded with a permanent status. However, our work suggests frequent movement is more typical, with workers often switching labor market categories. Although the percentage in each category may stay relatively stable over time, the actual people in those groups are frequently trading places, and that place-trading is a telling trait of this market.

In our study, we define "low wage" as a wage rate at or below the 30th percentile of the U.S. wage distribution. Although setting that cutoff is straightforward, identifying low-wage workers is not as easy. There really are no "low-wage workers," only low-wage jobs into which people are slotted during one or more periods of their lives. In our sample of over 7,700 people observed every other year from 1997 to 2013, 50% of those with wages are never low wage in the years we observe them while 50% are low-wage at least once during the sample years (Table 1, next page). The fact that a full 50% of the population marches through these markets at some point in their lives is a fact to keep in mind.

If we zero in on those who were ever low wage (i.e., low wage at least once), we find that even the people in this group spent only 42% of their earning years in low-wage jobs while 45% of their earning years were spent in higher-wage positions. As for those "stuck" in permanent low-wage status, we find that those cases are rare. Only 4.4% were low wage in all years. In fact, if we limit our focus to those with positive wages who are observed in at least five of the nine survey years, the percentage with permanent low-wage status drops to 1.2%. It is in this sense that most people defy categorization as low-wage workers. They simply work in low-wage jobs at some point in their work lives.

Although permanent low-wage status seems like a rarity according to this preliminary look, one might argue for broadening the focus to those who have either low or zero wage (LZW) in any given year. We explore that group next, and the results show why the panel analysis is so valuable. Of those between the ages of 19 and 64, only 30% are never LZW in the years we observe them, while 70% are LZW at least once during the sample years. This is not surprising. Among those with zero wages, we see people who are attending school (9%), people who are temporarily or permanently retired (13%), people reporting a permanent disability (15%), the unemployed (26%), workers on temporary layoff (3%), and a substantial number who report their activity as "keeping house" (32%). Omitting these people from the analysis would cloud our understanding, as their status can be a result of unattractive opportunities in the labor market. In fact, we observe movement back into employment even from retirement and "permanent" disability status. Keeping these people in the analysis for now, we find that those seemingly "stuck" in permanent LZW status make up 16% of the sample. Among those who were ***ever*** LZW, 23% are LZW in ***all*** years observed. Of course, this means that the vast majority are *not* permanently stuck in that category. If we focus on those observed in at least five of the nine survey years, the ***"permanently"*** LZW group shrinks to 17%, and the percentage *not* "stuck" rises to 83%. Again, most people who pass through the low-wage labor market defy categorization by permanent low- or zero-wage status. This is not to trivialize the consequences of low wages, especially for those who *are* trapped in that world. Instead, the point is that the low-wage job market is in constant motion, churning through a frequently changing workforce.

TABLE 1
Temporary Versus Permanent Low-Wage Status

Category		
Never low wage		49.8%
Ever low wage (at least once during the sample years)		50.2%
Percent of earning years in low wage jobs		41.5%
Percent of earning years with higher wages		44.7%
Low wage all years observed		4.4%
Observed at least 5 of the 9 survey years		1.2%
Ever low or zero wage (at least once during the sample years)		70.0%
Never low or zero wage		30.0%
Low or zero wage all years		16.4%
Observed at least 5 of the 9 survey years		12.0%
Of those ever low or zero wage		23.4%
If observed at least 5 of the 9 survey years		16.6%
Employment Status	**Zero Wage**	**Positive or Zero Wage**
Working	0.0%	76.9%
Temporarily laid off	3.0%	0.7%
Looking for work, unemployed	26.3%	4.8%
Retired	13.0%	4.5%
Permanently disabled	15.2%	3.4%
Housewife; keeping house	32.3%	8.0%
Student	8.5%	1.4%
Other	1.8%	0.4%

Source: Author's calculations based on extract drawn from the PSID for the survey years 1997 to 2013. All calculations use data weighted by the PSID family weights.

Labor Force Transitions in Low-Wage Markets for Workers of Prime Working Age

Labor force transitions are distinctly different for workers in low-wage jobs. If we examine transitions among four different categories of labor market status, we see this quite clearly. We limit this look to workers in their prime work years of age 25 to 54 in order to minimize interference from late schooling and early retirement, and we report their transitions for all survey years from 1997 to 2013. For those with wages above the 30th percentile prior to each survey year, a full 93% remain employed from one survey year to the next (Table 2, next page). That figure drops to 84% for workers starting from low-wage jobs prior to each survey and to 82% for people with a recent history of a zero-wage year.

The flip side of this story is the movement from employment to being unemployed or out of the labor force (OLF). For the higher-wage group, the percentage moving from employment to OLF or unemployment is relatively low. While 2.9% become unemployed, 3.4% exit the labor force. Such transitions are more than twice as high for workers starting from low-wage positions (15%) and nearly three times as high for those reporting a year of zero wages in the recent past (17%). Importantly, exits from employment into OLF status are highest, and substantially so (10.4%), for the latter group.

TABLE 2
Transitions Among Labor Force Categories by Wage Category

Table 2a: High Wage in Income Year t-3

Employment Status, Survey Year t-2	Employment Status, Survey Year t: Employed	Temporarily Laid Off	Unemployed	Out of the Labor Force	Total
Employed	93.1	0.6	2.9	3.4	100%
Temporarily Laid Off	68.0	7.2	9.1	15.7	100%
Unemployed	66.9	1.2	20.3	11.6	100%
Out of the Labor Force	48.9	0.7	7.5	42.9	100%

Table 2b: Low Wage in Income Year t-3

Employment Status, Survey Year t-2	Employment Status, Survey Year t: Employed	Temporarily Laid Off	Unemployed	Out of the Labor Force	Total
Employed	84.4	0.9	6.3	8.4	100%
Temporarily Laid Off	58.6	4.1	12.6	24.6	100%
Unemployed	57.4	1.5	26.9	14.2	100%
Out of the Labor Force	42.3	1.3	7.9	48.5	100%

Table 2c: Zero Wage in Income Year t-3

Employment Status, Survey Year t-2	Employment Status, Survey Year t: Employed	Temporarily Laid Off	Unemployed	Out of the Labor Force	Total
Employed	82.0	0.8	6.8	10.4	100%
Temporarily Laid Off	33.4	0.0	40.2	26.4	100%
Unemployed	41.7	1.2	29.9	27.2	100%
Out of the Labor Force	20.7	0.3	4.8	74.2	100%

Notes: Author's calculations from the PSID for survey years 1997-2013. Sample includes all persons ages 25–54.

Higher-wage workers fare better if placed on temporary layoff. For those who recently held a position paying above the 30th percentile, 68% move back into employment after a temporary layoff. The corresponding figure for those recently in a low-wage job is 59%. Recent higher-wage status also provides an advantage to the unemployed, with 67% reporting employment in the subsequent survey year, compared with 57% for those with a recent history of a low-wage job and 42% for those with a recent year without wages. For the last group, instead of finding employment, the majority of unemployed workers either remain unemployed or drop out of the labor force (57%). Remaining unemployed or dropping out of the labor force also characterizes 41% of unemployed workers who recently held a low-wage job. For those recently in higher-wage jobs, the corresponding figure drops to 32%.

Even leaving OLF status appears easier for those who held a higher-wage job before dropping out the labor force. Among those with higher wages prior to their OLF status, 49% are back at work by the subsequent survey year. For those starting from low-wage jobs prior to OLF status, the figure dips to 42%, while for those with a zero-wage year prior to the OLF status, only 21% move back to work by the next survey year. The figures for these 25- to 54-year-olds remain basically unchanged if we remove from the analysis persons reporting permanent disability, retired persons, and students.

Women in all three wage categories are much more likely than men to move to OLF status from employment, unemployment, or temporary layoff and to remain OLF once there (Appendix Table A3). Yet the likelihood of moving from these labor force categories into OLF status is largest for the women reporting low or no wages in the recent past. If recently in a low-wage job, 20% of unemployed women compared with 7% of unemployed men left the labor force after becoming unemployed. The figures jump to 33% and 20%, respectively, for women and men if their recent past includes a zero-wage year. The unemployed men, while less likely to exit the labor force, are more likely than the women to remain unemployed.

Movement between low- and higher-wage jobs is another common form of transition, and one that challenges use of the label "low-wage worker." Workers with wage rates below the 30th percentile often move up, suggesting that the job, rather than the worker, determines the wage status. Mobility of that type can result from upgrading and downgrading of hiring standards during business cycles. The pattern is one emphasized by macroeconomic theories of cyclical labor market mobility and is known for increasing the chances for women, workers of color and those with less education to gain a permanent foothold in markets that would otherwise be out of reach[4] (Okun 1981). Between any two survey years, 35% of low-wage women and 47% of low-wage men move from a wage that is at or below the 30th percentile to a wage that is above that cutoff (Table 3). That type of upward mobility is the least likely for low-wage women who are single heads of households. Yet, even for that group, 33% make such wage gains from one survey year to the next.

TABLE 3
Transitions Between Low- and High-Wage Jobs by Gender and Family Status

	Wage Status Relative to 30th Percentile, Year t		
Wage Status, Relative to 30th Percentile, Year t-2	**At or Below**	**Above**	
Women			
Low Wage	65.32	34.68	100%
Wage over 30th Percentile	12.55	87.45	100%
Men			
Low Wage	53.2	46.8	100%
Wage over 30th	8.09	91.91	100%
Single Women Who Head Households			
Low Wage	67.44	32.56	100%
Wage over 30th	15.72	84.28	100%
Wives			
Low Wage	63.91	36.09	100%
Wage over 30th	11.12	88.88	100%
Not Wives or Heads of Households			
Low Wage	61.54	38.46	100%
Wage over 30th	16.67	83.33	100%

Notes: Author's calculations from the PSID, survey years 1997-2013. Sample includes all persons ages 25–54.

We also note downward mobility. While only 8% of high-wage men drop from the higher- to the lower-wage category from one survey year to the next, 13% of higher-wage women experience such wage erosion. Again, wives fare better than single women who head households. While 11% of high-wage wives drop from higher- to lower-wage positions, 16% of high-wage single women who head households exhibit this type of downward mobility in wage status.

The churning in low-wage markets is also visible in the data on job tenure or "years with present employer." Accumulating multiple years with the same employer is much less likely for those in low-wage jobs. For all persons, average tenure is 5.9 years. That figure drops to 3.3 years for workers who are in low-wage jobs and to 2.2 years for workers who are in low-wage jobs 80% of the time. The higher volatility can also be characterized by the number of times people change their labor market status. Taking advantage of the long-period panel data, we focus here on those observed in all nine survey years between 1997 and 2013. While workers recently in higher-wage jobs change labor force status 11% of the time, those recently in low-wage jobs switch labor force status 18% of the time. The figures are higher for women than for men and highest (20%) for single women who head households.

The Great Recession and Long-Run Sequences of Employment Exits and Re-Entry

While the constant movement that we have documented is interesting in and of itself, we have yet to connect the observed movement to national or local shocks that frequently jolt low-wage markets. We do this now in two stages. First we explore the period of the Great Recession, tracking sequences of moves during the slump and ensuing recovery. Second, we use regression analysis to estimate the magnitude and statistical significance of links between market adjustments and increased labor market slack.

We explore sequences of moves starting in 2005, from well before the Great Recession; progressing through 2007 to 2009 when the recession gains full force; continuing into 2009 to 2011 when the recovery starts; and ending with 2011 to 2013, when the recovery gathers momentum. We consider all workers between the ages of 19 to 64 in order to include school and retirement reactions to the changing market conditions. In addition, we group workers by the wage reported in a year just prior to the survey years being explored. From 2005 to 2007, despite being a time of relative stability, the risk of moving from employment to unemployment is already higher for workers recently in low-wage jobs (4% versus 2%). More important, this risk more than doubles from 2007 to 2009, reaching 9% in that period and remaining elevated between 2009 to 2011 (7%) and 2011 to 2013 (6%). The rise for workers recently in higher-wage jobs is much more modest.

Moves from employment to retirement display a gradual upward trend from 2005 to 2013 for those recently in low-wage jobs and show a substantial jump, from 1.8% to 4.2%, in the depths of the recession for those recently reporting a zero-wage year. Some of these same workers re-enter the labor force at a later date.

Leaving employment for "keeping house" becomes less common during the recession, likely reflecting the greater reluctance to leave a job when economic uncertainty threatens livelihoods and savings. The fall-off in this move from employment to "keeping house" is most evident for those recently reporting a year of no wages. Transitions from employment to student status move in the opposite direction for this group. A rising percentage of those reporting a recent experience of low or zero wages move from employment to education as the job market worsens in 2007-2009. The same is true for those who begin the period unemployed. They show an increasing tendency to enter student status as the recession unfolds. We see signs that this pays off when we view sequences that include student-to-employment moves in the final years of our sample. By 2011-2013, we see a notable rise in the percentage entering employment from student status. Plus, those who had a recent year of zero wages appear to remain in school longer just after the recession. This is suggested by the increased figures, post-recession, in the "Student to Student" transition for this group.

To carefully document the sequences of moves driven by the Great Recession, we identified 35 possible paths from employment in 2007 to loss of employment in 2009 or 2011 and back to employment by 2013. This method shows that certain sequences are much more common than others for workers starting from low-wage jobs. Moving from employment in both 2007 and 2009, to temporary layoff in 2011, and back to employment in 2013 is less typical for men and women starting from low-wage jobs than for those starting the period in higher-wage positions (Appendix Table A4). A similar sequence, this time with employment in 2007–09 interrupted by unemployment in 2011, is much more common for all. Among men who start and end the period employed, over a fifth from both low- and high-wage jobs report this series of changes, a sequence that is less common for low-wage women. By far the most common sequence for those who start and end the period with a job is employment in 2007, followed by unemployment in 2009 with a return to employment in 2011 and 2013. Comparison of the two sequences shows that, for men starting from low-wage jobs, unemployment hits earlier. Reports of unemployment in two consecutive survey years describes yet another and more difficult path back to employment. A much larger percentage of men from low-wage jobs report this sequence of events (17%). The comparable figure for men recently in a better-paying job is much lower (6 %). In contrast, after employment ends, the women are more likely to report "keeping house" as their activity.

A small but interesting group consists of people who, after starting the period employed, move into retirement or disability status and then re-enter the labor force by 2013. If we combine all those who make temporary moves from employment into retirement or disability status and back to employment again, we find that more low-wage women (7.2%) than low- or higher-wage men (6%) make such recession-timed moves.[5]

A larger group of people drop out of the labor force during the recession to "keep house" or pursue education. We focus first on the moves from employment to education and back to employment again. A higher percentage of those from low-wage positions follow that path in the recession years. Summing over the high-unemployment years 2009-2011, we find 13% of women and 11% of men from low-wage jobs making such moves compared to 9% of higher-wage workers.

So far, we have focused on those who succeeded in returning to employment by 2013. Yet many paths end in less success, without a return to employment by 2013. Several patterns are worth noting. First, temporary layoff does not guarantee a return to employment, even for those in higher-wage positions. "Temporary" layoff is not always temporary. Second, even as the economy recovers, labor markets lag behind. Third, the lag in labor markets means that, despite staying employed in 2007 and 2009, 10% of the men from low-wage jobs report unemployment in the final two survey years, 2011 and 2013. That extended unemployment is again less common for women and for those who recently held a higher-wage job. Fourth, while unemployment in 2011 often ends in retirement by 2013, those starting from low-wage jobs are likely to move from unemployment in 2011 into disability status in 2013. This difference is understandable given the lower savings, smaller pensions, and greater physical demands typical of low-wage work. Women who recently held low-wage jobs are again more likely to follow a report of unemployment in 2011 with a report of "keeping house" in 2013. Fifth, moving from employment to student status in 2009 or 2011 is no guarantee of employment by 2013. Among those who are no longer in school in 2013, 7.1% of women and 6.3% of men from low-wage jobs have not returned to employment by 2013 even after ending their student status in 2009 or 2011. For those recently holding a better-paying job, ending student status, but failing to become employed, is much more unusual (2%).

When Hours Decline: Estimating the Variety and Scale of Adjustments in Low-Wage Markets

We now make the statistical links explicit by estimating the magnitude and significance of adjustments in low-wage markets when hours decline. Studies of labor markets often report estimates of the "elasticity of employment" with respect to specific economic shocks. In efforts to determine the meaning of those studies, some in the popular press interpret a negative elasticity with respect to changes in the minimum wage as jobs wiped out. Yet a decline in hours can be distributed in multiple ways, not all of which culminate in jobs eliminated. For example, a simple regression analysis of annual hours on state unemployment rates shows that a single percentage point increase in those rates causes annual hours to fall by 8% overall, 5% for workers who are low wage at least 50% of the time that we observe them and 17% for workers LZW at least 50% of the years observed (Table 4, next page). Some of those hours show up as reduced hours per week, which fall 4.0%, 3.3%, and 10%, respectively, for all persons and for the two wage categories. Other adjustments reflect fewer weeks worked per year, 4.4%, 2.6%, and 9.1%, respectively. By contrast, the probability of working (i.e., having positive hours or not) falls by about .0096, .0067 and .02, respectively, starting from a proportion with positive hours of 0.84 for the sample as a whole.[6] Though significant, those declines appear economically small. Likewise, the probability of losing all annual hours of work worsens by .004 for all persons and by .006 for those recently in a low-wage job.[7] Although not reported in Table 4, we also find evidence of offsetting ***increases*** in hours of overtime particularly for low-wage single women who head households.[8] The low-wage market thus has multiple ways to adjust to economic shocks, and lost hours of one type can be partially offset by increased hours elsewhere.

TABLE 4
Impacts of a Percentage Point Change in State Unemployment Rates
(between year t and year t+2, 1997-2013)

Dependent Variable	All Persons	Low Wage at Least 50% of the Time	Low or Zero Wage at Least 50% of the Time
Annual hours	-23.4 ***	–14.4**	–37.5 ***
ln(annual hours)	–8.1%***	–5.3%***	–17%***
Average hours per week	–0.44 ***	–0.40 ***	–0.93 ***
ln(average hours per week)	–4.4%***	–3.3%***	–9.6%***
Weeks per year	–0.48***	–0.17	–0.81***
ln(weeks per year)	–4.4%***	-2.6%**	–9.1%***
Tenure in years[a]	0.69 ***	0.54 ***	0.46 ***
ln(tenure)	11%***	12% ***	12%***
Annual overtime hours[b]	4.8 ***	6.3*	6
ln(overtime hours)	0.7%	2.6%	0.7%
Annual pay from extra jobs[c]	$68.70	$70.90	$70.00
Ln(pay from extra jobs)	6.47%	–1.48%	–1.47%
Probability of working[d]			
Linear probability estimates	–0.00961***	–0.00668***	–0.02137***
Probit estimates	–0.00666 ***	–0.00516***	–0.01509***
Probability of losing all hours[e]			
Linear probability estimates	0.00374***	0.00609***	0.000937
Probit estimates	0.0025***	0.00461***	0.00583

Notes: Based on data drawn from PSID, all persons ages 19–64, 1997 to 2013.
Calculations are panel estimates with individual and year fixed effects and data weighted by PSID's longitudinal family weights. Significance tests use clustered robust standard errors with clustering at the individual level.
a. Sample limited to wives and heads of households, with positive values for tenure.
b. Sample includes wives, heads of households, with positive values for overtime hours.
c. Sample limited to heads of households with positive values for pay from extra jobs.
d. Dependent variable equals 1 if hours are positive, zero if hours are zero.
e. Dependent variable equals 1 if change in hours from year t-2 to year t is negative and hours in year t are zero. Otherwise the dummy equals zero (zeros include those who kept some hours, gained hours, or had no change in hours).
Key for significance levels: *p<0.10; **p<0.05; ***p<0.01.

One might reasonably argue that, without controls for confounding factors, these estimates are less than precise and likely to be biased. Two points are relevant here: By controlling for personal characteristics, we implicitly assume that all of those personal traits are fixed. Yet one point of this study has been to suggest that markets and people often change and that markets can spur those changes in behavior. Second, while we do find that controls for personal characteristics alter the magnitude and significance of the estimates, this is perhaps the less notable finding. A fact that we find much more interesting is the role that tenure on the job (or what PSID calls "years with present employer") plays in influencing estimated effects by gender, family status and wage category (Table 5, next page). Our controls are the following: age, age-squared, high school education, college, number of children, age of the youngest child, whether white, and whether low or zero wage in a prior, but recent, survey year. In addition, we interact the dummy variable for the LZW category with the state unemployment rates in order to estimate the extra impact of unemployment on people with recent LZW status, and we show the results both with and without the tenure variable.

TABLE 5
The Interaction of Low-Wage Status with State Unemployment Rates
(between year t and year t+2, 1997–2013)

Dependent Variable = Annual Hours	**Male Household Heads**				**Female Household Heads**				**Wives**			
	model a		**model b**		**model a**		**model b**		**model a**		**model b**	
Unemployment rate, (UR)	-11.27	***	-13.11	***	-12.17	*	-18.31	***	–6.92	*	-10.71	***
(If low wage in t-2) x UR	-20.96	***	–12.12	**	–19.39	**	–7.80		–8.05		3.90	
If low wage in t-2	25.11		20.83		35.63		10.70		–119.21	***	–129.44	***
Age	98.74	***	73.50	***	94.80	***	73.90	***	82.16	***	52.01	***
Age-squared	–1.37	***	–1.12	***	–1.23	***	–1.05	***	–1.06	***	–0.78	***
High school	–78.26		–75.00		229.87	***	216.74	***	3.05		–9.39	
College	89.97	**	89.85	**	156.50	**	131.84	**	96.27	***	77.88	**
No. of children	–17.80	**	–6.59		–55.46	***	–44.50	**	–196.76	***	–176.52	***
Age youngest child	0.28		–0.91		1.12		1.97		12.62	***	12.86	***
White	35.28		20.86		–64.06		36.77		12.00		23.11	
Years with present employer			25.85	***			–62.06	***			37.00	***
Constant	520.01	***	944.50	***	–161.29		208.86		50.05		603.49	***
Number of Observations	35,729		35,521		15,356		15,288		29,968		29,819	

Notes: PSID, all household heads and wives, ages 19–64, 1997 to 2013. Calculations are panel estimates with individual and year fixed effects and data weighted by PSID's longitudinal family weights. Significance tests use clustered robust standard errors with clustering at the individual level.

* p<0.10; **p<0.05; ***p<0.01.

Here is what we find most telling. For both women and men, for male household heads, and for wives and single women who head household, the impact of recently holding a low-wage job significantly and substantially increases the impact of rising unemployment rates on the measures of hours discussed above. However, with addition of the tenure variable, the statistical significance disappears for all but the men, and even for them the magnitude drops dramatically. The meaning of this becomes clear when we substitute tenure as the dependent variable (Table 6, next page). For most workers, tenure increases when unemployment rates rise. Workers hold on to jobs longer when the economic future is uncertain, and employers hoard favored workers, protecting their investment in skilled workers when recessions hit. For the low-wage group, in contrast, we see a significant decline in tenure equal to a quarter of a year for each percentage point rise in the unemployment rate. While low-wage workers who have been with an employer for several years might ride out an economic shock, adjusting through reduced weeks per year or reduced hours per week, and offsetting with extra overtime, those with less than a quarter year of tenure will be out of luck, even with a single percentage point increase in the state's unemployment rate. If the rise is more substantial, those who have accumulated more tenure on the job will start to fare poorly.

TABLE 6

The Impact of Unemployment and Low-Wage Status on Tenure
(between year t and year t+2, 1997–2013)

Dependent Variable = Tenure	Male Heads of Households		Single Female Heads of Households		Wives	
State Unemployment rate, (UR)	0.08	**	0.17	***	0.10	***
(If low wage in t-2) x UR	–0.33	***	–0.30	***	–0.32	***
If low wage in t-2	0.09		0.57		0.23	
Age	0.95	***	0.56	***	0.83	***
Age-squared	–0.01		0.00	***	–0.01	***
High school	0.13		0.07		0.29	
College	0.06		0.65		0.58	**
No. of children	–0.43	***	–0.30	***	–0.55	***
Age youngest child	0.05	***	–0.03		–0.01	
White	0.73	*	–0.42		–0.24	
Constant	–16.25	***	–9.31	***	–15.29	***
N	35,521		15,288		29,819	

Notes: Based on sample from PSID of all household heads and wives, ages 19–64, for the period 1997 to 2013. Calculations are panel estimates with individual and year fixed effects and data weighted by PSID's longitudinal family weights. Significance tests use clustered robust standard errors with clustering at the individual level.

* $p<0.10$; ** $p<0.05$; *** $p<0.01$

Putting It All in a Larger Context

We have shown that permanent and uninterrupted low-wage work is less common than low-wage work interrupted by spells outside of that market. These interruptions include periods of nonparticipation, unemployment, education, home-making, and forays into better-paid work. Even if we limit our view to people who are observed in their prime working years and observed in nine of the survey years from 1997–2013, we still see this stop-start pattern of engagement with low-wage jobs. Those who were ever low-wage workers average about one third of their observed years in low-wage positions. These same individuals average about 16% of their observed years with zero wages. In fact, one could view the percentage of time in zero-wage status as a function of the low wages and poor work conditions that these jobs provide. Of course, a reaction to poor opportunities is only one way to interpret the data. The stop-start nature of these jobs also reflects the vulnerability of this market to bombardment from economic shocks.

One view that we question is that prime-age men are disappearing completely from participation in the labor market. Fixed snapshots of cross-section data contribute to this view. Panel data reveals that the men who make up that group of supposed non-participants are revolving in and out of the low-wage market. In other words, a different set of individuals makes up that non-participant group in any given time period. Instead of complete withdrawal from the labor market, as suggested by the cross-section data, these men seem to take turns dipping into the market and then leaving it, a behavior that may be perfectly rational given the working conditions that characterize most low-wage jobs. We observe a continual flow of people from low-wage to higher-wage jobs, from wage to no-wage activities, and back again, and that seems the more typical pattern in these labor markets. That observation belies the static "men-permanently-without-work" story that suggests something is wrong with the men themselves.

Our investigation also suggests the significance of an older story, one in which a rational response to poor job conditions is high job turnover and low attachment to jobs. According to that theory, primary sector jobs, with on-the-job training, skill accumulation, stable employment, job ladders, decent wages and fringe benefits, encourage attachment between employer and employee while secondary jobs, with little training or skill acquisition, no promise of stable employment, zero chance for advancement, low wages and few fringe benefits, discourage long-term attachment between employers and workers. In most versions of this theory, it is not the workers who bring poor traits to the jobs, but, instead, it is the typical design of the job that promotes high turnover, lack of investment in workers, and lack of attachment. Worker behavior is understood as a rational reaction to the poor conditions and lack of opportunity that these jobs create. An interesting policy possibility emerges from this perspective. As Zuberi creatively demonstrates in his book, "Differences that Matter," the same job task, even one performed for the same hotel chain, can offer dramatically different work conditions, benefit packages, wages, and opportunities if performed under vastly different policy regimes[9] (Zuberi 2006). For example, the hotel workers in Seattle in 2001 faced typical, secondary market conditions and poor job incentives while, just across the border in Vancouver, hotel workers for the same company had benefits and job conditions less characteristic of the secondary market. The picture of low-wage markets that emerges from our analysis aligns with this concept of using higher wages and pro-labor policies to transform secondary work into jobs that promote attachment, lead to lower turnover, and perhaps yield gains in productivity. That approach offers an intriguing possibility for addressing the traits of low-wage markets that we have documented here and for thinking in new ways about the possible consequences of minimum-wage hikes.

Appendix

TABLE A1
Variables: Descriptive Statistics

Variable	Definition	Mean	Std. Dev.	Min.	Max.
Age	Age in years	41.85	12.50	19	64
anHRS	Annual hours of work	1576.82	964.91	0	5824
anWKS	Weeks of work per year	37.86	19.15	0	52
AWH	Average weekly hours of work	34.35	19.10	0	112
Black	Proportion black	0.13	0.34	0	1
Children	Number of children in family	0.84	1.15	0	11
College	Dummy=1 if years of education is greater or =16	0.32	0.47	0	1
Ed	Education in years	13.35	2.80	0	17
ErnXTRA	Earnings from extra jobs (if positive)	$138.39	$269.26	0.01	$2,763
High school	Dummy=1 if years of education is greater or = 12	0.89	0.31	0	1
If low wage	Dummy=1 if person is low wage in prior survey year	0.24	0.42	0	1
LostALL	Dummy=1 if person lost all annual work hours	0.0553	0.2285	0	1
LRZ	Dummy=1 if person is low or zero wage prior year	0.3620	0.4806	0	1
Observed	Number times observed in the sample	7.13	2.64	1	9
OT	Overtime work hours (if positive)	160.16	199.56	1	2704
PosHRS	Dummy=1 if person has positive annual work hours	0.8367	0.3697	0	1
Tenure	Years with present employer	5.93	8.18	0	49
UR	State unemployment rate in year t prior to survey year	6.1	2.1	2.2	14.4
Wage	Wage rate in real dollars, base year=2015	$27.57	$36.35	$0.01	$1,316.41
Low wage	Wage rate below the 30th percentile	$8.39	$2.87	$0.01	$13.07
White	Proportion white	0.77	0.42	0	1
Youngest	Age of the youngest child	3.25	5.09	0	17

Source: Author's calculations from the Panel Study of Income Dynamics for survey years 1997 to 2013. All figures are weighted by PSID family weights.

TABLE A2
Weighted Observations by Gender, Age, Family Category

Category	Number	Percent
Total	93,877	100%
by Gender		
Male	44,272	47.16%
Female	49,605	52.84%
by Family Category		
Single Female Heads of Households	14,091	15.01%
Not Heads of Households or Wives	9,961	10.61%
Wives	31,282	33.32%
Husbands or Single Male Heads of Households	38,543	41.06%
by Age Category		
Young Workers (ages 19–24)	9828	10.47%
Prime Working Age (25–54)	65835	70.13%
Older Workers (ages 55–64)	18214	19.4%

Source: Author's extract from the Panel Study of Income Dynamics for survey years 1997 to 2013. All figures are weighted by PSID family weights.

TABLE A3
Transitions Among Labor Force Categories by Gender

a: High Wage in Income Year t-3

	Employment Status, Survey Year t				
Employment Status, Survey Year t-2	**Employed**	**Temporarily Laid Off**	**Unemployed**	**Out of the Labor Force**	**Total**
Employed					
Men	94.56	0.5	3.13	1.81	100%
Women	91.39	0.73	2.57	5.32	100
Temporily Laid Off					
Men	67.9	9.09	11.99	11.03	100
Women	68.06	5.54	6.6	19.8	100
Unemployed					
Men	65.65	1.23	25.71	7.4	100
Women	68.75	1.24	12.45	17.56	100
Out of the Labor Force					
Men	55.17	1.11	9.7	34.02	100
Women	46.49	0.51	6.67	46.33	100

(Table A3 continues, next page)

b: Low-Wage in Income Year t-3

Employment Status, Survey Year t-2	Employment Status, Survey Year t				
	Employed	Temporarily Laid Off	Unemployed	Out of the Labor Force	Total
Employed					
Men	87.01	1.01	8.11	3.87	100%
Women	82.87	0.88	5.23	11.02	100
Temporily Laid Off					
Men	62.16	10.13	19.4	8.31	100
Women	56.36	0.27	8.31	35.06	100
Unemployed					
Men	59.41	1.77	31.43	7.39	100
Women	55.68	1.33	22.96	20.03	100
Out of the Labor Force					
Men	45.4	1.07	11.06	42.48	100
Women	41.75	1.39	7.25	49.62	100

c: Zero-Wage in Income Year t-3

Employment Status, Survey Year t-2	Employment Status, Survey Year 7				
	Employed	Temporarily Laid Off	Unemployed	Out of the Labor Force	Total
Employed					
Men	87.73	1.07	7.01	4.2	100%
Women	76.68	0.54	6.55	16.23	100
Temporily Laid Off					
Men	51.56	0	31.12	17.33	100
Women	0.35	0	56.84	42.81	100
Unemployed					
Men	42.41	1.37	35.99	20.24	100
Women	41	1.11	24.62	33.27	100
Out of the Labor Force					
Men	21.61	0.02	6.94	71.42	100
Women	20.46	0.35	4.34	74.85	100

Notes: Author's calculations from the PSID, survey years 1997–2013. Sample includes all persons ages 25–54.

TABLE A4

The Great Recession and Long-Run Sequences of Labor Force Exits and Re-Entry

Sequences of Moves During the Great Recession and Recovery [a]				All Wage Groups	Low or Zero Wage in Previous Survey			Wage Above 30th Percentile in Previous Survey
2007	2009	2011	##		All	Women	Men	
E	E	T	E	3%	1.7%	2.4%	0.8%	4.5%
E	E	U	E	21.2	19.2	17.1	22.3	24.5
E	E	R	E	1.3	1.9	2.6	0.8	1.1
E	E	D	E	1.0	0.9	0.6	1.2	1.3
E	E	H	E	5.5	5.7	9.4	2.0	5.8
E	E	S	E	2.9	2.6	3.0	1.5	2.6
E	U	E	E	30.1	29.6	23.9	38.3	31.3
E	U	U	E	8.8	10.8	6.7	17.0	5.6
E	H	E	E	4.2	4.1	6.8	0	4.9
E	H	U	E	0.9	1.3	1.8	0.5	0.7
E	H	H	E	2.3	4.4	7.2	0	0.5
E	S	E	E	6.1	4.4	2.5	7.3	5.8
E	S	U	E	0.6	1.4	2.3	0	0
E	S	S	E	0.7	1.1	1.5	0.4	0.3
E	D	E	E	1.0	0.5	0.6	0.3	1.1
E	D	H	E	0.2	0.4	0.7	0	0
E	R	E	E	1.4	1.6	1.9	1.3	1.4
E	R	R	E	1.0	1.3	1.5	1.0	0.8
E	T	E	E	4.0	2.2	2.6	1.6	6.3

a. Sequences are defined in the key provided below.

Key: D=disabled; E=employed; H=keeping house; R=retired; S=student; T=temporarily laid off.

Source: Author's calculations from PSID, survey years 2007–2013 using PSID family weights.

Endnotes

[1] Eberstadt, Nicholas. 2016. *Men Without Work: America's Invisible Crisis.* West Conshohocken, PA: Templeton Press.

[2] For two comprehensive and competing perspectives on the existing literature see Belman, Dale, and Paul J. Wolfson. 2014. *What Does the Minimum Wage Do?* Kalamazoo, MI: W.E. Upjohn Institute for Employment Research and Neumark, David and William L. Wascher. 2008. *Minimum Wages.* Cambridge. MA: The MIT Press.

[3] This hypothesis is suggested by Hirsch, Barry T., Bruce E. Kaufman, and Tetyana Zelenska. 2015. "Minimum Wage Channels of Adjustment." *Industrial Relations,* Volume 54, Issue 2, pp. 199-239.

[4] Okun, Arthur. 1981. *Prices and Quantities: A Macroeconomic Analysis.* Washington, D.C: Brookings Institution Press.

[5] Note: Not all such moves are included in the table. The calculations are available from the author upon request.

[6] The Probit estimates are slightly lower than these linear probability estimates.

[7] The dependent variable is 1 if the change in hours from year t-2 to year t is negative and if all annual hours of work are lost. The variable is zero for those who kept some hours, gained hours, or had no change in hours. In comparison, the proportion of the sample as a whole losing all annual hours averages 0.0553 over the full period from 1997 to 2013. The estimates from the linear probability model and the Probit model differ substantially for the LZW group, ranging from 0.00094 from the linear probability model to 0.0058 from the Probit model.

[8] Those results are available from the author upon request.

[9] Zuberi, Dan. 2006. *Differences That Matter: Social Policy and the Working Poor in the United States and Canada.* Ithaca, NY: Cornell University.

References

Belman, Dale, and Paul J Wolfson. 2014. *What Does the Minimum Wage Do?* Kalamazoo, MI: W.E. Upjohn Institute for Employment Research.

Eberstadt, Nicholas. 2016. *Men Without Work: America's Invisible Crisis.* West Conshohocken, PA: Templeton Press.

Hirsch, Barry T., Bruce E. Kaufman, and Tetyana Zelenska. 2015. "Minimum Wage Channels of Adjustment." *Industrial Relations*, Volume 54, Issue 2, pp. 199-239.

Neumark, David, and William L. Wascher. 2008. *Minimum Wages.* Cambridge, MA: The MIT Press.

Okun, Arthur. 1981. *Prices and Quantities: A Macroeconomic Analysis.* Washington, D.C: Brookings Institution Press.

Zuberi, Dan. 2006. *Differences That Matter: Social Policy and the Working Poor in the United States and Canada.* Ithaca, NY: Cornell University.

IX. Toward a Secure Retirement: Challenges and Solutions

Retirement Income Is Inadequate for Most Americans

HEATHER GROB
Saint Martin's University

Typically, we think of retirement savings as being a three-legged stool: personal retirement savings, pensions, and Social Security. For most individuals, the lack of either of the first two leads to an overwhelming anxiety about the availability of the last. This session examines the viability of two very important parts of the stool: individuals' retirement savings and pensions.

The U.S. Government Accountability Office (GAO) study, "Retirement Security: Most Households Approaching Retirement Have Low Savings," is particularly helpful in asking whether or not vulnerable populations or people who are near retirement have enough savings, other than pensions. Here we find a story of "haves" and "have nots." Some very striking figures are that a full 41% of individuals in pre-retirement ages (between the ages of 55 and 64) have no other form of savings, and 85% have less than $25,000. The median is $104,000, which means that we have a terribly lopsided distribution of retirement savings. Older individuals with few retirement assets are also far less likely to have defined benefit pensions. Poverty risks to the "have nots" are great.

Income replacement is complicated, and the GAO does a good job of addressing methodology while reviewing seminal studies in the field of retirement income adequacy. By any measure, we can say that large portions of American workers are ill prepared for retirement. There is a huge disparity by household income, with the risks being much higher for low income: long-term care or medical costs needs could rapidly deplete what little savings they have.

The pension crisis is to a large degree, one that is born by women, as shown by the National Institute on Retirement Security (NIRS) study (Brown et al. 2016). While participation in pension plans are greater than they used to be, eligibility is still a problem. Systemic underfunding is a larger problem stemming from labor market segmentation along gender and race. For example, women are more likely to work in short term or part time employment and are therefore less likely to qualify for defined contribution plans than are men. And, for both men and women, the number of those offered and participating is still quite low, about 46%.

Of course, for those who have opportunities to invest, a very significant problem is the gender wage gap. The conclusion in the NIRS study is fairly glaring. Not only do women have fewer wages but they're also less likely to be able to retire, less likely to be able to enjoy their retirement, and more likely to fall into poverty in old age. Relative to men, women who are 65 or over are 80% more likely to fall into poverty. Women aged 75 through 79 are *three times* more likely to fall into poverty than are men.

The problem worsens for people of color as the labor market further segregates Latinas and black women into occupations that are less likely to allow for retirement savings. Brown et al. (2016) also indicate that when offered retirement plans, women of color are more likely to participate than are men of color. More research would be helpful indicating why this is the case: is it that men of color don't expect to reach retirement or have other plans for their income? Interestingly, composition of income between men and women is very similar, but the dollar amounts contributed are not. People of color therefore find themselves highly reliant on Social Security as their sole source of retirement savings. Unfortunately, Social Security is not keeping up with the Elder Index either.

Author's address: Saint Martin's University, Old Main 369, 5000 Abbey Way SE, Lacey, WA 98503

Not surprisingly, women who worked in jobs as nurses, teachers, or in government administration were better off because they were more likely to have a saving grace: collectively bargained, defined benefit plans.

The NIRS report could further highlight particular issues dealing with complicated issues of marital status and changing definitions of marriage. The report highlights that women who were separated or never married are more likely to be working past 65 than are men, and more than widowed or divorcees, and they lack access to defined benefit and defined contribution plans.

From a policy standpoint, we could take a cue from behavioral economics, deaSling with the psychology of loss aversion and perceived risks. Why don't people save? There are many reasons, of course: ability, opportunity, wishful thinking, attitudes, and so on. The forthcoming report by The Pew Charitable Trusts will be very helpful understanding what employers and individuals like and do not like about automatic enrollment in individual retirement accounts (IRA). They use a very clever methodology to get to specifics. The good news is that both workers and employers of small- to medium-sized businesses generally like the idea of automatic IRA enrollment. Employers want to help their employees save, and, when they are unable to offer their own plan, this is an attractive alternative. When asked if they would offer their own plan or take part in the government's plan, roughly half of employers said they would start their own. It is interesting to note that many oppose government intervention while they seem to support financial intermediation.

The NIRS report also discusses many possible solutions including: eligibility changes, portable savings accounts, exclusion changes under the Employee Retirement Income Security Act of 1974, Social Security benefit enhancements, caregiver support, and automatic IRA, the state retirement and savers plan or new defined benefit plans may also help.

To add, there are also systemic risks that are due to reliance on employee benefit system: job loss, access to health insurance through one's job, and so on. More recently, the further "fissuring" of employment relations through involuntary self-employment, independent contracting, subcontracting and franchising was documented by Weil (2014). We are dealing with a system that has been "fracked": employment benefits are very fragile. As I mentioned at last year's meetings on the viability of multi-employer plans, the amount of lost income not just to individuals but throughout our economy of these perpetuating and worsening conditions of retirement adequacy should be of great concern to all.

These studies are cross sections of the retirement savings world and provide ample room for thought and further study. My main question is whether or not retirement income should be based on employment at all. Our labor markets are changing in ways that threaten the basic system of relying on employment benefits: we will have to think of creative very solutions for people to be able to save what they can, when they can.

References

Brown, Jennifer, Nari Rhee, Joelle Saad-Lessler, and Diane Oakley. 2016. *Shortchanged in Retirement: Continuing Challenges to Women's Financial Future.* Washington, DC: National Institute on Retirement Security. http://bit.ly/2k6xoG4

Scott, John, Andrew Blevins, Theron Guzoto, Alison Shelton, and Sarah Spell. "Employer and Employee Reactions to State-Level Retirement Savings initiatives for Private Sector Workers." Unpublished paper. The Pew Charitable Trusts.

U.S. Government Accountability Office. 2015. "Retirement Security: Most Households Approaching Retirement Have Low Savings: Report to the Ranking Member, Subcommittee on Primary Health and Retirement Security, Committee on Health, Education, Labor and Pensions." U.S. Senate GAO-15-419. http://bit.ly/2k6N0JX

Weil, David. 2014. *The Fissured Workplace: Why Work Became So Bad for So Many and What Can Be Done to Improve It.* Cambridge, MA: Harvard University Press.

X. Rethinking Pension Fund Activism for Employment Equity and Capital Stewardship

Mortgage Fraud, the Great Recession, and Pension Fund Activism

THOMAS HERNDON
Loyola Marymount University

Losses in private mortgage backed securities (MBS) were at the epicenter of the 2007-2008 financial crisis, causing the failure of institutions invested in them or derivatives based on them. I argue that these losses revealed substantial problems with asymmetric information in MBS which created perverse incentives for abuses, including outright fraud. These losses were particularly severe for pension funds, who have had limited ability to recover losses through ex-post litigation. I argue that pension funds have unique tools, including shareholder activism and access to capital, which would be valuable for activism aimed at preventing these abuses in the future.

Introduction

Losses in private mortgage backed securities (MBS) were at the heart of the financial crisis of 2007-2008. The failure of the mortgages underlying these securities caused substantial losses for the institutions directly invested in them, such as pension funds, as well as loss of wealth for the communities in foreclosures occurred. The purpose of this paper is to discuss how pension fund activism can improve capital stewardship by helping to address the significant problems with the regulation of financial institutions revealed by the failure of private MBS. This paper will argue that losses in private MBS revealed substantial problems with asymmetric information throughout the structure of private mortgage securitization. These problems allowed financial industry insiders to use private information to profit at the expense of users of the financial system, often through outright fraud. This paper will then discuss how pension funds have tools unique to them, such as shareholder rights and access to capital, that would allow them to make a substantial contribution to activism aimed at improving protection from deception for all those who rely on the financial system.

This paper is organized into two sections. The first section will describe how asymmetric information in the origination, distribution, and servicing portions of the private mortgage securitization supply chain allowed financial industry insiders to use deception and outright fraud to profit at the expense of borrowers, savers, and shareholders in financial institutions. This section will describe how executives of institutions originating loans to be privately securitized had perverse incentives to systemically conceal borrower risk in an effort to increase loan volume at the expense of quality. The section will then describe how sellers of private MBS concealed the fraudulent and negligent origination practices from those who purchased these securities. Finally, the section will discuss how servicer fee compensation created a conflict of interest that prevented loss mitigation for delinquent loans, and increased costs of foreclosure for investors. These problems caused a historic loss of wealth for borrowers subjected to fraudulent loans, investors in securities based on these loans, and shareholders of fraudulent financial institutions who either lost their investment, or were left to pay the fine. These losses show that pension funds have a common interest in promoting increased protection for all users of financial institutions. The second section will describe how pension funds have unique tools available that would allow them to make a substantial contribution to efforts to activism aimed at preventing financial sector fraud, and recovering from the damage caused by it. This section will discuss two potential avenues for pension fund activism. The first is shareholder activism to prevent fraudulent or

Author's address: Loyola Marymount University, University Hall, Suite 4218, 1 LMU Drive, Los Angeles, CA 90045 abusive financial practices. I argue that pension funds could use the tools available to them as shareholders in financial institutions to eliminate perverse incentives in executive compensation, develop stronger internal controls against fraud, monitor private information, and work with consumer protection groups, regulatory agencies, and the media to hold financial institutions accountable for abusive practices. Second, I argue that pension funds could help working-class communities and communities of color recover from the Great Recession through working with non-profit financial institutions to purchase distressed mortgages, and restructuring the debt to allow families to remain in their home.

To be sure, the outcome of activism and organizing is based on a conflictual process, and hence is always uncertain. We should be sanguine about the ability of these efforts to completely eliminate fraud, or otherwise ensure the financial system serves social goals. Still, I argue pension funds possess powerful tools that could make meaningful contributions to activism. If we are fortunate, this may result in an improvement to the large shortcomings of current regulation of financial institutions.

Mortgage Fraud in Origination, Distribution, and Servicing During the Great Recession

The market for private MBS grew dramatically from 2002-2007, with the total outstanding balance of mortgages increasing from roughly $1 trillion to $2.7 trillion (Herndon, 2016*a*). This provided a large increase in the supply of credit for mortgages which drove the housing bubble (Mian and Sufi, 2014; Griffin and Maturana, 2014). As the housing bubble collapsed, losses in these securities were at the epicenter of the financial crisis of 2007-2008. A large body of academic research (Black, 2013; Crotty, 2009; Griffin and Maturana, 2016; Herndon, 2016*a*,*b*; Jiang, Nelson and Vytlacil, 2014; Mian and Sufi, 2015; Piskorski, Seru and Witkin, 2015; Taub, 2014), government reports (FCIC, 2011; FBI, 2007), court records, and popular accounts (Hudson, 2010; Dayen, 2016) has now shown that a significant portion of losses in these securities was caused by mortgage fraud all along the originate to distribute supply chain. For example, as early as 2004 the FBI warned of an epidemic of mortgage fraud that could cause a financial crisis (Black, 2013). Additionally, the Financial Crisis Inquiry Commission (FCIC) concluded that a systemic breakdown in accountability and ethics was an essential contributor to the crisis. The report cites the dramatic increase in the filing of suspicious activity reports with the FBI as evidence of the widespread increase in mortgage fraud in the years leading up to the crisis. The filing of suspicious activity reports grew twentyfold between 1995-2005 and then doubled again between 2005-2007 (FCIC, 2011). This section will review this literature to describe the severe problems in the origination, distribution, and servicing portions of the private securitization supply chain.

Fraud at Origination

Mortgage fraud at origination occurred when loan officers and underwriters used a variety of techniques to falsify borrower financial information such as appraisal value inflation, unreported second liens, income overstatement, and misreported owner occupancy status. This was done to conceal borrower leverage and risk to qualify borrowers for loans larger than they would otherwise be able to obtain. This section will begin by describing how fraud originated within the industry, rather than by dishonest borrowers. The section will then describe how short-term compensation for financial industry executives created perverse incentives for fraud.

The direct falsification of borrower financial information was largely committed by loan officers and underwriters within the industry, who coached borrowers on the specific ways to falsify their information, rather than by borrowers who defrauded otherwise honest lenders. For example, based on investigations and fraud reports, the FBI found that 80% of fraud cases involved collusion or collaboration with industry insiders (FBI, 2007). Interviews in lawsuit documents with loan officers or underwriters also described coaching borrowers. One loan officer who originally worked at subprime originator Argent, but then was employed at Wells Fargo, said that, "[t]he loan officers were stretching the truth. They would say [to the borrower], 'You need to make this much.' So, of course, the borrower would say, 'Ok, I make that much'."[1] Even worse, a loan officer from Ameriquest, Omar Khan, explicitly described deceiving borrowers who were not comfortable with falsifying their information. He stated, "Every closing was a bait and switch, because you could never get them to the table if you were honest," and further elaborated that "there were instances where the borrower felt uncomfortable about signing the stated income letter, because they didn't want to lie, and the stated income letter would be filled out later on by the processing staff."[2] Perhaps most infamously, workers at one Ameriquest branch dubbed their break room the "Art Department" because it contained all the tools needed to falsify documents (Hudson, 2010).

Interviews with loan officers and underwriters also described immense pressure coming from top-level executives to falsify documents in order to expand loan volume at the expense of loan quality, as well as penalties for refusing to do so. For example, one confidential witness was employed as an Underwriter at Wells Fargo Home Mortgage in San Bernardino from 2002 to 2005, and Senior Underwriter from May 2005 to April 2006. This underwriter described Wells Fargo as a, "loan producing machine." They stated that, "[Managers] always said that we didn't have to approve loans we didn't want to approve, but if you didn't do them you wouldn't be around very long. We knew what we had to do to keep our jobs." They elaborated that, "sometimes it felt like I was in sales, because they wanted production, period."[3] In addition to firing employees who refused to originate risky loans, those who called attention to risky or fraudulent practices were also punished. For example, Washington Mutual CEO Kerry Killinger hired and fired nine different compliance officers from 2000-2007 (Taub, 2014).

An obvious question is why executives would pressure loan officers, underwriters, and compliance officers to facilitate fraudulent loans, when it was these employees' *raison d'etre* to ensure that such practices did not occur. Fraud at origination occurred because executives at these institutions had perverse incentives to increase short-term profits based on the volume of loans originated, rather than the quality of loans. This was because executives were able to receive large bonus compensation for short-term gains, for example through stock options, that were not required to be paid back if the firm went bankrupt.[4] Fraud was particularly useful for increasing short-term revenues because riskier loans had higher closing costs and interest rates. This allowed originators to report high short-term fee revenue that could be extracted before losses occurred. This pattern of fraud is also similar to that which occurred during the S&L crisis (Black, 2013).

Another incentive to originate risky loans is that many of these loans could be sold for packaging into MBS for a higher price than safer loans (Taub, 2014). This has led to calls for financial regulation that requires originating institutions to have "skin in the game" by holding a portion of the mortgages they originate on their balance sheets. However, these skin-in-the-game regulations would likely not have stopped fraud at origination. Indeed, many of the originators did hold a large portion of the toxic loans in their portfolio, and went bankrupt as a result. These institutions in fact had substantial skin in the game, which caused their failure. However, their executives did not. The pattern of extreme executive compensation, despite the failure of their firms, could reasonably be described as "looting." Looting occurs when owners or executives have limited liability for a firm, and maximize short-term payouts at the expense of the long run health of their firm, resulting in bankruptcy. Looting has been described as bankruptcy for profit (Akerlof and Romer, 1993). Preventing looting would likely have required increased monitoring of institutions, limits to extreme compensation packages, and criminal prosecution of top executives (Black, 2013).

Fraud in Distribution

Fraud also occurred in the distribution phase of the supply chain because sellers of MBS concealed these fraudulent origination practices from investors in order to make the securities marketable. The sale of loans that were originated with fraudulent practices, or simply negligent underwriting, typically violated market regulations and contractual obligations. These rules require the accurate disclosure of loan quality; however, these practices obviously were not disclosed.[5] The basic issue underlying fraud in distribution was succinctly summarized in a recent ruling by District Judge Denise Cote,

> "This case is complex from almost any angle, but at its core there is a single, simple question. Did the defendants accurately describe the home mortgages in the Offering Documents for the securities they sold that were backed by those mortgages? Following trial, the answer to that question is clear. The Offering Documents did not correctly describe the mortgage loans. *The magnitude of falsity, conservatively measured, is enormous.* Given the magnitude of falsity, it is perhaps not

> surprising that in defending this lawsuit defendants did not opt to prove that the statements in the Offering Documents were truthful" [emphasis added].

A recent body of empirical research has also estimated the incidence of mortgage quality misrepresentation in private MBS. For example, using conservative measures Griffin and Maturana (2016) find that 48% of loans that were privately securitized contain at least one of three relatively easy to quantify forms of fraud: appraisal inflation, unreported second liens, and misreported owner occupancy status. They find that loans with one of these forms of fraud were 51% more likely to become delinquent.

My research has also shown that this fraud was also particularly costly to investors in these securities. Total losses to foreclosure in the private-label MBS was roughly $500 billion dollars from 2008-2012. My research found that loans that lacked documentation of income, which were so notoriously fraudulent that they were known colloquially within the industry as "Liar's Loans," accounted for $350 billion of these losses. Liar's loans were 25% more likely to default than a control group of full documentation loans, and lost roughly $20,000 more in foreclosure. Therefore, the higher than expected losses to investors due to this single form of fraud account for roughly one-fifth of total losses to foreclosure, or $100 billion (Herndon, 2016*a*).

In contrast to the problems with originating institutions that could reasonably be described as looting, the problems in the market for securities based on these loans are more accurately described as a "market for lemons." The term "lemon" refers to a car that is poor quality, or more generally, to any product that is poor quality. A market for lemons is a market where good and bad quality products are sold, but where the buyers cannot know beforehand whether they are buying a good or bad product. In these markets, bad products tend to push out good products because good and bad products must sell at the same price (Akerlof, 1970). Over the course of the housing bubble, it is clear that bad practices in this market had pushed out good practices because a "significant degree of misrepresentation exists *across all* reputable intermediaries involved in the sale of mortgages," [emphasis in original] (Piskorski, Seru and Witkin, 2015).

Litigation is the most direct method for defrauded investors to recover losses and deter future fraudulent activities. To this end, diverse institutions that bought fraudulent MBS, including pension funds, government agencies, and mortgage insurers, have initiated a large amount of litigation against all major sellers of MBS.[7] To be sure, there have been several notable settlements. The latest tally of recoveries I was able to locate is from 2014, and puts the total amount recovered at roughly $100 billion (Levitin, 2014). Since then, there have been several additional settlements, which would raise the total amount recovered to perhaps around $130-$140 billion. However, I argue that the outcome of these settlements shows the limits of private investors to recover damages or create a credible deterrent against future fraud.

First, the settlement total shows that in general, purchasers of MBS have not been able to enforce their contractual rights and recover damages from those that defrauded them. On the face of it, the aggregate settlement amount of roughly $130-$140 billion is much less than the $500 billion lost to foreclosures in the private-label RMBS market from 2007-2012. Additionally, from the perspective of any single party defrauded in the purchase of RMBS, these lawsuits required large commitments of time and money for an uncertain outcome. This suggests that all parties faced significant limits to their ability to recover damages.

More significant for pension funds, the distribution of settlements shows that private investors were able to recover much less than any other party. Private MBS litigation settlements only account for roughly 2% of the total settlement amount as of 2014, or roughly $1.67 billion. Indeed, private MBS litigation has only been able to recover roughly 4% of the amount that GSEs were able to, and one-quarter of what monoline insurers have been able to recover (Levitin, 2014). To be sure, some public pension funds, such as CalPERs have recently recovered large amounts as part of DoJ settlements. CalPERs was able to recover roughly $250 million from Bank of America, $261 million from JP Morgan, $88 million from Citigroup, $100 million from Moody's, and $125 million from S&P.[8] However, private pensions and other investors have had difficulty recovering the full amount of damages. Indeed, at roughly $800 million, the settlements received by CalPERs alone are roughly 50% of the total recovery by private MBS legislation as of 2014.

Second, this body of litigation has also shown the limits of any party to impose substantive deterrents for fraudulent behavior on financial crises. First, while I have not been able to locate data on the profits generated by the financial sector in the sale of MBS, the financial penalties imposed seem far too small to make the costs of engaging in fraud higher than the payoff for a financial institution. Additionally, there has been a noticeable lack of prison sentences for executives who oversaw MBS fraud, or any other form of fraudulent behavior for the activities that led to the financial crisis. In stark contrast, there were over 1000 felony convictions during the S&L crisis (Black, 2013). Without criminal prosecution of the actual executives responsible for fraud, it is unlikely fraud will be deterred. Fines that are imposed on institutions will be paid for by shareholders and employees, rather than the executives who committed the actual abuses.

Fraud in Servicing

Servicer misconduct also negatively affected investors by increasing the number of foreclosures that occurred and increasing the costs of foreclosure. Servicers often charged borrowers arbitrary fees and misapplied payments so that they could charge delinquency fees. This prevented delinquent borrowers from curing, and even pushed borrowers who never missed payments into foreclosure. Once in foreclosure, it was in servicers' financial interest to impose arbitrary fees that would be recovered out of foreclosure proceeds prior to any revenue given to investors. For example, in December 2013 one of the largest servicers, Ocwen, settled a complaint with the Consumer Financial Protection Bureau and attorney generals from 49 states for $2 billion. CFPB director, Richard Cordray, stated that, "Ocwen took advantage of borrowers at every stage of the process." The complaint documented how Ocwen "took advantage of homeowners with servicing shortcuts and unauthorized fees," "deceived consumers about foreclosure alternatives and improperly denied loan modications," and "engaged in illegal foreclosure practices."

At the root of servicer misconduct was a conflict of interest based on the servicer's cost-plus compensation structure. Servicer compensation is not aligned with the investors interest in maximizing the net present value of the loan. Instead, the servicer's choice of modification or foreclosure, and type of modification, is based on the incentives in their own compensation structure. Servicers receive three main types of income: a fixed-rate fee based on the unpaid principal balance of a loan; float income from the period in which the servicer receives monthly payments but has not remitted them to the trust; and ancillary fees. The main types of ancillary fees include delinquency fees and reimbursement for costs associated with foreclosure, such as property maintenance fees, title search fees, process serving fees, appraisal fees, other legal fees, or any of a number of other fees. There is no effective oversight of the reasonableness of these fees, and servicers are able to be reimbursed for these fees out of the proceeds of the foreclosure sale prior to any revenue being given to investors (Levitin and Twomey, 2011; Thompson, 2011; COP, 2009).

Lack of oversight of fees charged in foreclosure caused two main problems for investors during the waves of foreclosures that followed the collapse of the housing bubble. First, these fees can be quite lucrative and create an incentive to foreclose, even when it is in the investor's best interest to modify, because modification is costly. Modification is costly for three reasons. First, modifications require substantial labor costs such as re-underwriting the loan. Second, if the modification reduces monthly payments through reducing the unpaid principal balance, the servicer loses its fixed-rate fee. Third, servicers must advance missed payments while the loan is delinquent. They can recoup these advances in cases of foreclosure or if the loan becomes current, but not in many types of modifications.

In contrast to the costs associated with modifications, the fees associated with managing delinquency and foreclosure can be quite lucrative. For example, analysis of Ocwen showed that late fees and loan collection fees made up 18% of its revenue in 2008 (Thompson, 2011). There can also be an incentive to keep a borrower delinquent so that the servicer can receive revenue from delinquency fees, until the cost of financing advances outweighs the revenue received from the fees. This has been described as keeping the borrower in a default fee "sweatbox" (Levitin and Twomey, 2011). Essentially, the servicer's choice between "modification and foreclosure is a choice between limited fixed-price income and a cost-plus contract arrangement with no oversight of either the costs or the plus components," (COP, 2009). Even worse for the investor, this cost-plus structure creates an incentive to foreclose in a more costly manner than less, because servicer compensation is positively related to costs and has the senior claim on foreclosure sale revenue. Cost-plus compensation is typically banned from government contracts due to these perverse incentives (Levitin and Twomey, 2011; COP, 2009).

The second problem created by this compensation structure is that it provides incentives for servicers to choose types of modifications that promote their own interests, even if these modifications have a higher re-default rate and hence do not promote the investor's interests. For example, reducing monthly payments through principal reduction has been shown to be the most effective form of modification at preventing re-defaults (Haughwout, Okah and Tracy, 2009; Goodman et al., 2012). However, servicers are disincentivized to perform principal reduction because it reduces the amount of revenue they receive from their fixed-rate servicer fee, which is assessed on the unpaid principal balance of the loan. In contrast, servicers prefer modifications that increase the unpaid principal balance of the loan through capitalizing missed interest payments and fees because this increases the revenue from their fixed-rate fee. But these modifications that increase borrower indebtedness have higher re-default rates, which result in costly foreclosure for investors. Providing unsustainable modifications designed to re-default can also be a source of profit for servicers, because they can receive the lucrative foreclosure fees described above (Thompson, 2011; COP, 2009).

An obvious question is what is preventing market competition from correcting the principal-agent problem by creating incentives for "good" servicers who can meet the needs of investors? Market competition is unlikely to self-correct the misalignment of incentives because of investors in these securities lack the ability to monitor servicers, and the mechanism to fix abuses. Investors cannot effectively monitor servicers because they typically lack the information to do so. In general, investors do not have access to the detailed loan-level data necessary to evaluate the re-underwriting of modifications. Additionally, investors often lack the mechanism to address abuses when detected due to collective action problems. Investors faced two main collective action problems for addressing problems. First, many pooling and servicing agreements for private MBS had collective action clauses requiring a super majority of investors to amend any contractual terms. However, there were typically large numbers of geographically dispersed investors party to most of the major securitizations. Second, the investors often had different interests regarding the type of loan modification they would desire because they received compensation based on different parts of the cash flow, such as principal or interest payments. Therefore, some modifications would be favorable to some subset of investors, while wiping out a different subset of investors. These information and collective action problems effectively undermined investors' ability to perform meaningful oversight of servicers (Levitin and Twomey, 2011).

The findings in my research are also consistent with the reports of servicer conflict of interest. I found that foreclosures were much more frequent than modifications, with 88% more foreclosures occurring. These foreclosures were also incredibly costly to investors, on average losing between 45%–65% of the original balance. I also found that the overwhelming number of modifications favored servicers' interests over investors' by increasing debt. Modifications that resulted in net reduction in debt were incredibly rare, with only 5% of modifications reducing net debt. Indeed, modifications in this market resulted in a total net increase to borrower debt balances of $20 billion from 2008-2014. The amount of debt added per modification also grew from 2010-2014, roughly doubling from 5.6% to 11.3% of the original balance, or from $16,000-$26,000. Additionally, the growth in debt added per modification is consistent with increased fees assessed by servicers, but not increased missed interest payments, because missed interest payments per modification was constant from 2010-2014 (Herndon, 2016*b*).

What Is to Be Done?

The discovery of the problems in the previous section caused private investors to abandon the market for private MBS. New originations in this market disappeared entirely in 2009, and have since only existed at a low level. This caused to total outstanding balance of this market to shrinking to less than $1 trillion in 2014, from its $2.7 trillion peak in 2007. Currently, the mortgage market is roughly 80% government supported, with GSE's guaranteeing roughly 60% of mortgages, and the FHA insuring another 20% (Levitin and Wachter, 2013). However, going forward almost all reform proposals for the secondary market envision a substantially increased, if not exclusive, role for private institutions.

A common theme of the description fraud in the previous section was that the problems of asymmetric information that allowed insiders to defraud outsiders negatively affected all users of the financial sector. Financial institution insiders used their access to private information to profit at the expense of borrowers, savers, and shareholders in their institutions. Therefore, going forward pension funds have a common interest with all other users of the financial sector in eliminating these problems. This will be particularly more important if the private role in the secondary market is substantially increased in the near future. This section will describe how pension funds have unique tools that could enable pension fund activism to make a significant contribution to preventing fraud in the future, and recovering from the damage. Based on the tools available to pension funds, this section proposes two areas for activism that could potentially be fruitful: shareholder activism and debt relief.

Shareholder Activism and Corporate Governance

In this section, I describe how tools available to pension funds in their position as a shareholder could make significant contributions to fraud prevention. This includes preventing fraud that directly affects pension funds as shareholders in looted institutions, as well as fraud and abusive practices affecting users of services offered by the financial institution in which a pension fund is invested. In particular, pension funds could help to eliminate fraud through preventing perverse incentives in executive compensation packages, strengthening internal controls for fraud prevention such as compliance managers, and direct monitoring of fraudulent or abusive activities. Direct monitoring could also work in tandem with consumer protection groups, regulatory agencies, and the media to hold financial institutions accountable for abusive practices. To be sure, the ability of shareholder activism to achieve its goals relies on the outcomes of conflicts between parties, and thus is always uncertain. That being said, these particular tools should provide real points of leverage to be used in this conflict.

As we saw in the discussion of fraud in the origination of mortgages, perverse incentives for executives drove them to loot their companies. Looting occurs when executives have limited liability and the ability to extract large payments from their institutions, especially compensation such as stock options that do not have to be paid back in event of insolvency. This gives them the incentive to generate large short-term cash flows based on fraudulent behavior and extreme leverage, which can then be extracted from the institution before the losses inevitably come due (Black, 2013). If shareholders are able to be effectively organized, they have some formal power to monitor executive compensation and limit opportunities for extracting short-term cash flows. A simple mechanism for this would be "clawback" clauses that require executives to repay compensation in the event of firm failure. At the most extreme, this may require organizing to elect different members to the board of directors of the corporation. To be fair, limiting excessive executive compensation might prove to be extremely difficult. For example, there is some evidence to show that proxy resolutions to limit executive compensation have had less success than other resolutions (Daily, Dalton, and Rajagopalan, 2003). Still, if shareholders could successfully organize to limit of the ability of executives to extract payments, this would make a substantial contribution to fraud prevention.

Next, shareholder activism can strengthen internal controls to prevent fraud. Successful looting of financial institutions by executives requires them to use their power to disarm internal controls such as auditors and compliance managers. The financial crisis produced numerous reports from auditors and compliance managers who were fired for reporting fraud. For example, Kerry Killinger from Washington Mutual went through nine separate compliance managers from 2000-2007. Perhaps most ironically, an auditor named Ed Parker from Ameriquest earned the nickname of "Darth Vader" from the loan origination staff through his attempts to prevent origination of fraudulent loans. Instead of being promoted, he was fired (Taub, 2014). Fraud would not have been able to remain concealed or cause the same extent of damage had these whistleblowers' efforts received the support of shareholders or the board, rather than punishment. Without defeating these internal controls, top management will not be able to defraud borrowers, savers, and shareholders in their institutions.

Shareholders in corporations also often have greater access to private information than consumer protection groups or often regulatory agencies do. Direct monitoring by shareholders can play a large role in ensuring that the general public is not defrauded because successful fraud relies on concealing private information (Daily, Dalton, and Rajagopalan, 2003). Shareholders could possibly work in tandem with consumer protection groups, regulatory agencies, and the media. Consumer protection groups and regulatory agencies can be early warning groups who sound the alarm for abusive practices. Shareholders could use their access to inside information to confirm these practices. They could then work with the media and regulatory agencies, such as the Consumer Financial Protection Bureau (CFPB), to hold the executives responsible accountable for these abuses.

Debt Relief?

Pension fund activism could also speed recovery in working-class communities and communities of color through offering debt relief. I argue that pension funds should work with local non-profit financial institutions to buy distressed mortgages, and modify these mortgages to allow families to remain in their homes. Essentially, this proposal is for pension funds to help provide debt relief similar to the New Deal era Homeowner Loan Corporation (HOLC). At the height of the Great Depression in 1933, roughly one-half of the mortgages in the country were in default, and 10% were in foreclosure. To address this crisis, the HOLC bought up defaulted mortgages, wrote down the negative equity in these loans, and restructured the terms of the mortgages to create a more stable structure. In its first year, the HOLC received applications from 40% of all mortgage holders, and refinanced half of them (Levitin and Wachter, 2013). The HOLC is widely regarded as being highly successful, and I argue that pension funds could work with local non-profit institutions to provide similar debt relief. This would help prevent further foreclosures, and stimulate local aggregate demand through facilitating household deleveraging.

Currently, the largest government program for selling distressed loans has been the Distressed Asset Stabilization Program (DASP), which sells distressed loans insured by the FHA. In 2010, HUD began a pilot test for this program, and formally launched it in 2012. In 2014, Freddie Mac began to test pilot programs for distressed loan sales, with Fannie Mae following suit in 2015. As of 2016, these programs have sold roughly 100,000 distressed mortgages, with an unpaid principal balance of $18 billion, across 175 different pools. The distressed mortgages in the DASP program also overwhelmingly come from working-class and minority communities, with 84% of mortgage notes in the DASP program coming from communities with the share of minorities above the national median (Edelman, Zonta and Rawal, 2016). Therefore, debt relief would be well suited for helping these communities deleverage and recover. Currently, there are roughly 800,000 distressed loans guaranteed either by the FHA or the GSEs, so there is still ample room to increase the scale of purchases (Edelman, Gordon and Desai, 2014; Goodman et al., 2016). To date, the distressed mortgage market has been dominated by for profit financial institutions such as private equity firms. The two largest are Lone Star Funds, a private equity group which has purchased 23% of DASP loans, and Bayview Asset Management, which has purchased roughly 20% of DASP loans and is funded by Blackstone Private Equity (Goldstein, 2015).

However, similar to the description of servicer misconduct in the previous section, there have been significant consumer protection complaints lodged against these for profit financial institutions for refusing to modify loans. For example, borrowers have reported being in the process of negotiating a modification prior to sale of the distressed loan, but having the new servicer disregard the ongoing negotiations and initiate foreclosure. Additionally, servicers have been criticized for refusing to offer modification with principal reduction. HUD had sold the loans at a discount with the intent that private buyers would grant principal reduction to borrowers; however, this has rarely occurred. For example, Fitch Ratings reviewed loan modifications done by Calibre Home Loans, the lead servicer for Loan Star Funds, during the first half of 2015. Fitch Ratings was unable to locate even a single instance of a completed modification with permanent principal reduction. Instead, modifications included reductions in payments or only required interest payments for a temporary period of up to 5 years. At the end, any deferred unpaid principal or unpaid interest were added back to the principal balance of the loan, resulting in higher payments than prior to modification. Therefore, these modifications only added temporary relief, before often leaving borrowers more in debt than before. In a letter to HUD, the refusal to grant sustainable modifications in favor of foreclosures led Massachusetts Representative Michael Capuano (D) to remark that HUD sales, "may turn out to be an efficient new mechanism for increasing evictions," (Goldstein, 2015).

However, non-profit financial institutions have had a better track record of working to keep families in their homes. Several well-established community development financial institutions, which are non-profit financial institutions with a social mission, have also participated in HUD sales. For example, Hogar Hispano Inc. (HHI), founded by the National Council of La Raza, purchases delinquent mortgages and works with families to keep them in their home. HHI has already helped homeowners reduce $4 million in principal through modifications (Dreier and Sen, 2015). Additionally, Boston Community Capital has worked with the community organization Springfield No One Leaves to provide debt relief to homeowners (Kinney, 2013).

While non-profits have shown better performance at keeping families in their homes, to date they have only been able to purchase roughly 2% of loans from DASP (Edelman, Gordon, and Desai, 2014). To increase their participation, in 2015 HUD created new rules for the DASP bidding to give non-profits a "first-look," or the first option to purchase vacant properties, and took steps to create a non-profit only auction (Edelman, Zonta and Rawal, 2016).

I argue that pension funds could play a useful role in partnering with CDFIs or other non-profit institutions by providing the funding necessary to substantially increase their scale of participation in this market. Unfortunately, to date some public pensions have been funding the private equity institutions in this market, such as Lone Star (Goldstein, 2015). However, funding institutions with better loss mitigation practices could greatly help working-class families and communities of color. The recovery in these communities is still tepid, and thus vulnerable to increased foreclosures and predatory servicer practices. Debt relief would allow pension funds to prevent abusive practices in these communities, and stimulate recovery.

Conclusion

This paper has described how the failure of private MBS at the core of the 2007-2008 financial crisis revealed substantial agency problems associated with asymmetric information, including widespread fraud. Financial reform since the crisis, such as the restrictions on mortgage origination enacted by the CFPB, has helped to mitigate the worst of the excesses. However, the perverse incentives generated by asymmetric information have not entirely been eliminated. Moreover, there has been a noticeable lack of criminal prosecution of the executives who committed fraud to serve as a deterrent. This suggests that mortgage fraud, or other forms of abuse by the financial services industry, could continue to be a persistent problem negatively affect borrowers, savers, and shareholders in financial institutions. This would be especially dangerous if private participation in MBS increases to its pre-crisis levels, as is proposed in most secondary market reforms. Therefore, defrauded parties have a common interest in organizing to enact significant changes to financial regulation. Using the unique tools available to pension funds, I hope they will be able to make a substantial contribution to activism aimed at preventing fraud by the financial services industry in the future.

Endnotes

[1]Interview with confidential witness from *General Retirement System of the City of Detroit v. Wells Fargo et al.*, 2009.

[2]*National Credit Union Administration Board v. Wells Fargo Bank, National Association*, 2014.

[3]Interview with confidential witness from *General Retirement System of the City of Detroit v. Wells Fargo et al.*, 2009.

[4]Perverse incentives due to extreme bonus compensation were not limited to this market. They were a consistent feature of the expansion of the financial system following deregulation (Crotty, 2009).

[5]The typical offering documents included prospectus supplements that described the quality of collateral underlying the securities. These documents tended to include boilerplate language such as "Wells Fargo Bank's underwriting standards are applied by or on behalf of the Wells Fargo Bank to evaluate the applicant's credit standing and the ability to repay the loan, as well as the value and adequacy of the mortgaged properties collateral" [*General Retirement System of the City of Detroit v. Wells Fargo et al.*, 2009]. If the trustee discovered a breach of these representations and warranties, such as falsification of borrower financial characteristics, violations of assurances that loans were originated following proper underwriting standards, or that the appraisal value for the collateral was inflated, the "trustee must notify the appropriate parties and take steps to enforce the responsible parties obligation to cure, substitute, or repurchase the defective mortgage loans" [*National Credit Union Administration Board v. Wells Fargo Bank, National Association*, 2014]. It should be noted that origination practices that could be argued were simply negligent or dubious, but did not involve outright falsification, were still fraudulent in distribution because they violated the representations made in offering documents.

[6]From ruling in *Federal Housing Finance Agency v. Nomura Holding America*, May 11, 2015. The FHFA sued 16 trustees for misrepresentations made in offering documents and prospectuses for securities sold to Fannie Mae and Freddie Mac. All but Nomura and Royal Bank Scotland settled out of court, and the court ruled against these institutions in trial on May 11, 2015. http://nyti.ms/2kAGs5u [June 26, 2015].

[7]An older list of 58 lawsuits filed between 2008-2012 can be found in the appendix to Piskorski, Seru, and Witkin (2015). However, this list is not exhaustive, as the 2009 class action lawsuit used in this paper was not on the list (*General Retirement System of the City of Detroit v. Wells Fargo et al.*, 2009). In addition, several similar lawsuits have been filed for violations of the False Claims Act or the Financial Institutions Reform, Recovery and Enforcement Act (FIRREA), for actions such as misrepresenting the quality of loans to entities that insured these loans.

[8]CalPERs settlement totals can be found at http://bit.ly/2kAzFZN

References

Akerlof, George. 1970. "The Market for "Lemons": Quality Uncertainty and the Market Mechanism." *The Quarterly Journal of Economics,* 488–500.

Akerlof, George, and Paul Romer. 1993. "Looting: the Economic Underworld of Bankruptcy for Profit." Brookings Papers on Economic Activity, 1–73.

Black, William K. 2013. *The Best Way to Rob a Bank is to Own One: How Corporate Executives and Politicians Looted the S&L Industry.* University of Texas Press.

COP. 2009. "Foreclosure Crisis: Working Towards a Solution." Congressional Oversight Panel.

Crotty, James R. 2009. "The Bonus Driven Rainmaker Financial Firm: How These Firms Enrich Top Employees, Destroy Shareholder Value and Create Systemic Financial Instability." Political Economy Research Institute Working Paper.

Daily, Catherine M, Dan R Dalton, and Nandini Rajagopalan. 2003. "Governance Through Ownership: Centuries of Practice, Decades of Research." *Academy of Management Journal*, 46(2): 151–158.

Dayen, David. 2016. *Chain of Title.* The New Press.

Dreier, Peter, and Aditi Sen. 2015. "Hedge Funds: The Ultimate Absentee Landlords." *American Prospect.*

Edelman, Sarah, Julia Gordon, and Aashna Desai. 2014. "Is the FHA Distressed Asset Stabilization Program Meeting Its Goals?" Center for American Progress.

Edelman, Sarah, Michela Zonta, and Shiv Rawal. 2016. "Protecting Communities on the Road to Recovery." Center for American Progress.

FBI. 2007. "Financial Institution Fraud and Failure Report Fiscal Years 2006 and 2007."

FCIC. 2011. "The Financial Crisis Inquiry Commission Report: Final Report of the National Commission on the Causes of the Financial and Economic Crisis in the United States." Financial Crisis Inquiry Commission.

Goldstein, Matthew. 2015. "As Banks Retreat, Private Equity Rushes to Buy Troubled Home Mortgages." *New York Times DealBook.*

Goodman, Laurie, Erik Cribbs, Eric Stein, Don Mullen, and Julia Gordon. 2016. "HUD and GSE Nonperforming Loan Sales: Are Further Improvements Necessary?" Urban Institute Sunset Seminar.

Goodman, Laurie S, Lidan Yang, Roger Ashworth, and Brian Landy. 2012. "Modification Effectiveness: The Private Label Experience and Their Public Policy Implications." Washington, DC: Pew Commissioned Research Paper, 15.

Griffin, John M, and Gonzalo Maturana. 2014. "Did Dubious Mortgage Origination Practices Distort House Prices?" Available at SSRN 2485308.

Griffin, John M, and Gonzalo Maturana. 2016. "Who Facilitated Misreporting in Securitized Loans?" Review of Financial Studies.

Haughwout, Andrew, Ebiere Okah, and Joseph Tracy. 2009. "Second Chances: Subprime Mortgage Modification and Re-Default."

Herndon, Thomas. 2016a. "Liar's Loans, Mortgage Fraud, and the Great Recession." Dissertation at University of Massachusetts Amherst.

Herndon, Thomas. 2016b. "Punishment or Forgiveness? Loan Modifications in Private Label Mortgage Backed Securities from 2008-2014." Dissertation at the University of Massachusetts Amherst.

Hudson, Michael W. 2010. The Monster: How a Gang of Predatory Lenders and Wall Street Bankers Fleeced America—and Spawned a Global Crisis. Macmillan.

Jiang, Wei, Ashlyn Aiko Nelson, and Edward Vytlacil. 2014. "Liar's loan? Effects of origination channel and information falsification on mortgage delinquency." *The Review of Economics and Statistics.*

Kinney, Jim. 2013. "Homeowners Call for Mortgage Help at 'Springfield No One Leaves' Advocacy Group." *Springfield Republican.*

Levitin, Adam. 2014. "MBS Settlements Following the Money." *Credit Slips.*

Levitin, Adam J., and Susan M. Wachter. 2013. "The Public Option in Housing Finance." *UC Davis Law Review*, 46: 1111–1655.

Levitin, Adam J., and Tara Twomey. 2011. "Mortgage Servicing." *Yale Journal on Regulation*, 28(1).

Mian, Atif, and Amir Sufi. 2014. *House of Debt: How They (and You) Caused the Great Recession, and How We Can Prevent It from Happening Again.* University of Chicago Press.

Mian, Atif R., and Amir Sufi. 2015. "Fraudulent Income Overstatement on Mortgage Applications during the Credit Expansion of 2002 to 2005." National Bureau of Economic Research Working Paper 20947.

Piskorski, Tomasz, Amit Seru, and James Witkin. 2015. "Asset Quality Misrepresentation by Financial Intermediaries: Evidence from the RMBS Market." *The Journal of Finance*, 70(6): 2635–2678.

Taub, Jennifer. 2014. *Other People's Houses.* Yale University Press.

Thompson, Diane. 2011. "Foreclosing Modifications: How Servicer Incentives Discourage Loan Modifications." *Washington Law Review.*

XI. The Economics of Prevailing Wage Laws

Prevailing Wage Laws, School Construction Costs, and Side-By-Side Bids

KEVIN DUNCAN
Colorado State University–Pueblo

School districts in Maryland may ask contractors for two bids for the same construction project. One bid requires the payment of prevailing wages while the other does not. These side-by-side bids provide a unique opportunity to examine how contractor bid behavior affects the measured cost of the wage policy. Results indicate that the gap between the two bids decreases as bid competition and contractor bid experience increases. The disparity in side-by-side bids is also influenced by a contractor's eagerness to win a project. Additional analysis illustrates how the average bid gap of 9.9% disappears under particular bid behaviors and outcomes.

Introduction

Prevailing wage laws establish location and job-specific minimum wage and benefit rates for construction workers employed on public works projects. These standards apply to construction funded by the federal government, to building activity financed by 29 state governments, and by numerous municipalities.[1] Regardless of the jurisdiction, the purpose of the wage and benefit floor is to prevent large government projects from distorting local compens-ation standards.[2] Large projects may attract contractors from areas where wages are lower with competition between these low-wage, out-of-area builders and local establishments depressing area rates. The floor allows all contractors to compete without affecting wage and benefit rates that are determined in local construction labor markets. While research has examined the impact of the wage policy on local economic activity, safety and training in the construction industry, the racial composition of the construction labor force, and the provision of health and retirements benefits for construction workers, the public policy debate has centered on the impact of the wage floor on the cost of public construction (Duncan, Lantsberg, and Manzo 2015; Azari-Rad 2005; Bilginsoy 2005; Bellman 2005; Kessler and Katz 2001; and Waddoups 2005).

This paper contributes to the literature and the policy debate by exploiting unique school construction bid data from Maryland. School districts in this state may request contractors to submit two bids for the same project: one bid requires the payment of prevailing wages; the other bid does not. These side-by-side bid data allow for an examination of the level of bid competition, accumulated contractor bid experience, and contractor motivation to win a project on the relative cost of projects covered by the wage policy. The results of the study provide insight into the cost impact of the wage requirements as well as illustrate how contractor bid behavior evolves and responds to prevailing wage requirements. The remainder of this paper is organized as follows. The next section contains a survey of the existing literature on prevailing wage regulations and school construction costs as well as a description on how the data used in this study differ from the information that is typically available. The data and the statistical model are described in more detail in the following sections. The paper concludes with a discussion of results and implications for future research addressing the cost implications of prevailing wage regulations.

Author's address: Hasan School of Business, University of Colorado–Pueblo, 2200 Bonforte Boulevard, Pueblo, CO 81001-4901

Previous Research on Prevailing Wage Laws and School Construction Costs

Most of the studies examining the cost effect of prevailing wage regulations focus on school construction as these projects are relatively uniform and numerous.[3] Many of the school studies use project-level data obtained from Dodge Data & Analytics. This organization collects and distributes project bid information to the construction industry. Dodge reports the winning bid for a project, but does not include change orders that determine final (total) project costs.[4] This is a uniform practice across all data used in this literature where the winning, low bid is the measure of total construction costs. The Dodge data also contain information on project location and bid letting date that allow researchers to determine if prevailing wage regulations apply. Other detailed project-level data include measures of project size (square feet and number of stories), whether the project is new or an addition, and framing and flooring type, etc.

Azari-Rad, Philips, and Prus (2002 and 2003) use Dodge data to examine school construction across the U.S. during the 1990s and fail to find any statistically significant evidence that schools built in states with prevailing wage laws are more costly. Philips (2014) examines new school construction in Kentucky, Michigan, and Ohio when these states enacted, suspended, or repealed prevailing wage policies in the 1990s and finds that there is no statistically significant difference in average square foot costs associated with fluctuations in state-level wage policies. In an analysis of Maryland school construction, Prus (1999) finds that schools built in counties with prevailing wage requirements are no more expensive than the cost of comparable facilities built in counties that do not have the wage policy. Both of the studies by Prus and Philips utilize Dodge data. On the other hand, Vincent and Monkkonen (2010) also use Dodge data to examine school construction across the U.S. between 1995 and 2004 and find a prevailing wage cost effect ranging from 8% to 13%.

Several other studies use data similar to Dodge to examine the effect of the introduction of prevailing wage requirements on school construction in British Columbia.[5] This wage policy is similar to several strong state-level policies in the U.S.[6] Bilginsoy and Philips (2000) use the CanaData and find that public school bid costs under the wage policy did not differ in terms of statistical significance from the bids of public schools built before the introduction of the prevailing wage requirement. Duncan, Philips, and Prus (2014) examine the effect of British Columbia's prevailing wage standard by including a control group of private school projects. This difference-in-differences analysis indicates that before the introduction of the prevailing wage policy, the cost of building public schools was approximately 40% more expensive than the costs of comparable private schools. The differential between public and private school construction cost did not change after the wage policy was introduced. These authors have also used the British Columbian example to study the effect of prevailing wage laws on the productivity and efficiency of construction (see Duncan, Philips, and Prus 2012, 2009, and 2006). They find that prior to the introduction of the wage legislation, public school projects were 16% to 19% smaller, in terms of square feet, than comparable private structures (given the same project expenditure). This size differential did not change after the policy was in effect. These results suggest that prevailing wage requirements do not alter labor or other input utilization in a way that significantly affects the relative size of covered and uncovered projects. The authors also find that average total efficiency for public school construction is 94.6%. Average efficiency for projects covered by the introductory stage of British Columbia's construction wage legislation was 86.6%. By the time of the expansion of the policy 17 months later, the average efficiency of covered projects increased to 99.8%. These findings suggest that the introduction of prevailing wage laws disrupted construction efficiency. However, in a relatively short period of time, the construction industry adjusted to wage requirements by improving overall construction efficiency in a way that is consistent with stable total costs. A similar pattern was observed with respect to cost efficiency. While these studies utilize different sample configurations and statistical methods, they uniformly fail to find evidence that prevailing wages increase construction costs.

Atalah (2013a, 2013b) uses data obtained from the Ohio School Facilities Commission (hereinafter, OSFC) to test the hypothesis that prevailing wages increase school construction costs in Ohio. These data are limited to information that identifies the school district, participating contractors, all bid submissions for a project, and project size. The advantage of these data is that there are over 8,000 bids in the OSFC data set. With this information, Atalah is able to compare bids submitted by contractors who are signatories to collective bargaining agreements and to those submitted by "open shop" contractors. While schools were exempted from Ohio's prevailing wage law in 1997, union rates are the prevailing rates for other construction funded by the State of Ohio.[7] Consequently, Atalah's union-nonunion comparison is an indirect test of the impact of prevailing wage and benefit rates, omitting any other unique administrative costs associated with the policy.

Results from the first study (Atalah 2013a) indicate that the lowest (winning) bid costs per square foot for projects paying union wages are no more expensive than projects paying open shop rates. The exception is projects in the southern region of the state where bid costs per square foot are 51% lower for construction based on *union* rates. This difference is statistically significant at the 0.0005 level. While Atalah's first study examines the consequences of prevailing wage laws by comparing projects completed by union or nonunion workers, the second study (Atalah 2031b) compares the lowest bid costs by trade (plumbing, electrical, etc.) and union status. These results indicate that bid costs per square foot were higher for five of 18 (27.8%) of the trades involved in school construction that paid union rates. There were no statistically significant differences in bid costs per square foot for 72.2% (13/18) of the trades, regardless of union status.

Keller and Hartman (2001) use project cost data provided by the Pennsylvania Department of Education, applicable prevailing wage rates, and total compensation rates from a large nonunion contractor to examine the effect of Pennsylvania's prevailing wage requirement on school construction costs. By substituting nonunion wages for prevailing wage rates and adjusting for labor costs as a percent of total construction costs, these authors find that prevailing wages add 2.25% to the cost of building public schools. A shortcoming of the method used by Keller and Hartman is that their comparison of prevailing and open shop wage rates ignores the changes in labor productivity and utilization that take place when wages change in the construction industry. For example, Blankenau and Cassou (2011) report that the use of skilled and unskilled workers in the construction industry is sensitive to wage rates. Skilled workers are defined as those with more than a high school degree while unskilled workers have less than a high school education with the elasticity of substitution between these two grades of labor equal to approximately 9.0. Additionally, Balistreri, McDonald, and Wong (2003) find that capital equipment replaces labor when construction wages increase, though the elasticities of substitution between capital and labor are inelastic in the short and long-run. Taken together, the results of these studies indicate that labor productivity and utilization change with wage rates in the construction industry. The method used in the study by Keller and Hartman does not take these changes into consideration when calculating the cost impact of prevailing wages. As a consequence, the estimate reported in this study is too high.

This survey of the literature indicates that the preponderance of research fails to find a statistically significant prevailing wage cost effect. One reason why prevailing wages may not affect construction costs is that labor costs (wages and benefits) are typically a low percent of total construction costs. According to data from the *Economic Census of Construction*, labor costs (wages and benefits) represent about 23% of total construction costs for the entire U.S. construction industry in 2012 (U.S. Bureau of the Census 2012).

While most of the previous studies utilize data that include controls for project size and local economic conditions, it is also important to consider the influence of contractor incentives and bid behavior on the relative cost of projects covered by prevailing wage requirements. Contractors, particularly nonunion establishments who pay rates below prevailing levels, may adjust their bids on prevailing wage projects as they become more familiar with the requirements. Bids may also change with the level of bid competition, or eagerness to win a project. These issues have been largely ignored in the literature.[8] The side-by-side bid data for school construction in Maryland provide an opportunity to explore these issues when two bids are submitted by the same contractor for the same project, under equal local economic and market conditions. The side-by-side bids arise from the characteristics and provisions of the prevailing wage standard in Maryland.

Maryland's Prevailing Wage Policy, Side-By-Side Bids, and Contractor Bid Behavior

Prevailing wage rates for construction projects receiving funding from the State of Maryland are determined for all 23 counties and the City of Baltimore (State of Maryland, *Compliance and Frequently Asked Questions*). Minimum rates for projects covered by Maryland's prevailing wage regulation are determined by the following process: the prevailing wage rate is the rate paid to 50% or more of local workers in a detailed job classification. If fewer than 50% of local workers in a classification receive the same wage, the prevailing wage is the rate paid to at least 40% of the local workers in the classification. If fewer than 40% of local workers in the same job classification earn the same wage, the prevailing wage rate is the average wage, weighted by the number of workers receiving different wage rates. Between 2000 and 2014, prevailing wage requirements in Maryland applied to school construction projects with a value of at least $500,000 and when state funding was 50% or more of project construction costs. As of July 1, 2014, prevailing wages are required on projects with a value of at least $500,000 and when state funding is 25% or more of total construction costs (Maryland General Assembly, "*Prevailing Wage Law*").

School districts have the choice of opting out of prevailing wage requirements by accepting less than 25% in state funding (or less than 50% prior to July 2014).[9] When projects are expected to be close to either the $500,000 value threshold or to the minimum state funding contributions, school districts may ask contractors to submit two bids for the same project: one based on the payment of prevailing wages with the other ignoring this minimum wage requirement. These side-by-side bids allow a school district to determine which pay schedule is most advantageous by comparing the decrease in state funding to the bid-cost savings associated with avoiding the payment of prevailing wages. For example, if the side-by-side bids of the lowest submissions indicate a project cost savings of 20% by opting out of the wage policy, and if state funding for the project decreases by 10% if the wage regulations are avoided, it is practical for the school district to opt out of wage policy coverage.

Based on an examination of 266 side-by-side bids for 67 separate school construction projects, the Public School Construction Program found that, on average, bids based on prevailing wage rates were 11.7% higher than bids without prevailing wages. This cost impact is based on the comparison of all bids including the lowest bid for projects built between January 2012 and December 2015. This gap persists when only low bids are considered. For example, for the subset of roof replacement projects there were a total of 83 bids on 17 roofing projects between 2012 and 2015. The average gap between prevailing wage bids and bids that were not based on the payment of prevailing wages is 9.67%. The gap between the 17 lowest bids is 9.10%. The result obtained from the analysis of side-by-side bids is viewed as "incontrovertible evidence" that prevailing wages increase construction costs (Public School Construction Program, "*The Cost of School Construction*" 2015). Side-by-Side bids have been used elsewhere as evidence that prevailing wage laws increase construction costs. The Westlake City School District in Ohio required contractors to submit two bids, one subject to prevailing wage requirements and one bid exempt from the wage policy. An examination of these side-by-side bids suggests an overall construction cost savings of 5.8% without prevailing wages (Legislative Service Commission 2002).

The evidence based on the side-by-side comparisons is at variance with earlier research of Maryland schools. As described above, Prus (1999) finds no statistically significant cost difference in schools built in counties with and without prevailing wage requirements. An important difference is that Prus examines an array of school projects (new construction and renovations) while the side-by-side analysis is based on projects that are close to the project value and state funding thresholds. This is a critical distinction that influences contractor incentives, the disparity in side-by-side bids, and the implied cost estimate of prevailing wages. When school districts request side-by-side bids, they are sending a signal to contractors that some state funding may be sacrificed *if* significant savings can be promised by avoiding the payment of prevailing wages. Under these circumstances, contractors, particularly nonunion contractors, have an incentive to inflate estimates on prevailing wage bids.

To illustrate, consider a project with one nonunion bidder. Without any competition, both bids, with and without the payment of prevailing wages will be inflated. If this contractor wishes to avoid the payment of prevailing wage rates and other requirements of the policy including the submission of certified payrolls, apprenticeship registration, arranging benefits that meet prevailing standards, and other administrative responsibilities, the bid based on the payment of prevailing wages will be particularly inflated.[10] Expanding this concept to a more realistic setting with multiple bidders suggests that when bid competition is low and the likelihood of winning is relatively high, the difference in side-by-side bids may be relatively large. A tacit or collusive agreement to increase disparity in side-by-side bids may be made between contractors when bid competition is low. This type of arrangement is in the best interest of all nonunion contractors bidding on projects requesting two submissions and may be considered self-reinforcing to some extent. However, in a more competitive situation, the disparity in side-by-side bids may collapse as the likelihood of winning decreases and uncertainty over how other bidders will behave increases.[11]

Contractor experience with bidding on prevailing wage projects, as well as the dual-bid format, may also influence the gap in bids. Those who are new to prevailing wage projects may have greater uncertainty regarding all of the attendant requirements and regulations associated with the wage policy. As a consequence, less experienced contractors may pad these bids accordingly. As experience with this bidding format and the wage policy increases, contractors may reduce the disparity in bids that do and do not require the payment of prevailing wages. This suggests that relatively new bidders will have larger differences in side-by-side bids and that the gap between bids will decrease with accumulated bid experience.[12]

When a contractor is motivated to win a project, regardless of whether prevailing wages are required, it is likely that differences in side-by-side bids are reduced. This outcome may be observed during the peak bid season. For the counties and projects (roof replacements) examined in this study, 41% of all projects are open to bidding in March with 48% of all bids submitted during this peak month. It is likely that contractors who are very eager to win projects during the peak season submit low bids regardless of the payment of prevailing wages. Several other factors such as a backlog of unfinished work or the desire to work with a particular owner may also influence a contractor's motivation to win a project.[13] When a nonunion contractor is not eager to win, both bids may be higher with the bid based on prevailing wages being particularly high. Under these conditions, a contractor's bid may also be less competitive and finish with a higher ranking/place. This illustration suggests that if a contractor is highly motivated to win a bid, regardless of prevailing wage coverage, it is expected that the bid ranking will be lower as well as the disparity is side-by-side bids.

The policy change in 2014 that lowered the threshold for prevailing wage coverage to school projects receiving 25% of funding from the state may also affect the behavior of contractors and their side-by-side bids. According to information reported by the Department of Legislative Services, this policy change made virtually all K-12 projects funded by the State of Maryland eligible for the payment of prevailing wages that exceeded the $500,000 value threshold (Maryland General Assembly, "*Prevailing Wage Law*"). Under these conditions, nonunion contractors participating in projects requesting side-by-side bids may have responded to expanded prevailing wage coverage by inflating bids based on prevailing wages if they wished to avoid the requirements of the wage policy. This explanation suggests that the disparity in side-by-side bids will be larger after the July 1 policy change.

Side-By-Side Bid Data and Results

Data for the study were obtained from the Public School Construction Program, Interagency Committee on School Construction, Board of Public Works, State of Maryland. From January 2012 to December 2015, the Public School Construction Program collected 266 side-by-side bids for 67 school construction projects completed throughout the state. These projects largely consist of renovation work involving a variety of trades and tasks such as carpentry, concrete, demolition, drywall, electrical, flooring, HVAC, masonry, and roofing, etc. Roof replacement projects are selected for this study due to the relative homogeneity of these types of projects and the relatively large number of projects and bids. Over the period, there were 83 side-by-side bids by 18 different contractors on 17 roof replacement projects located in Carroll, Frederick, Howard, and Washington counties. Since 75 of these bids were submitted by ten contractors who participated in at least two projects between 2012 and 2015, an unbalanced panel of nonunion contractors was created for the statistical analysis.[14]

Table 1 includes data on the lowest and highest differences in contractor side-by-side bids. To illustrate, consider Contractor #1. In one of these bids submitted by this contractor, the difference between the prevailing wage bid and the bid without prevailing wages was as low as 5.3%. In another bid by this same contractor, the difference in side-by-side bids was as high as 30.1%. There is considerable variation between contractors. This is evidenced by the average difference in bids (see column 4 in Table 1). For example, the average bid difference for Contractor #1 is 12.7% and 5.5% for Contractor #2. The variation between contractors is also revealed by the range in lowest and highest differences. Contractor #5 submitted at least one bid where there was no difference between the prevailing wage and non-prevailing wage bid (where the lowest bid difference is 0.0%). On the other hand, Contractor #6 had one bid where the difference was as high as 42.1% (see highest bid difference for #6). The averages for the 75 bids included in the study indicate a mean low difference in side-by-side bids of 3.9%, a mean high of 20.5%, and an overall average gap in the two bids of 10.2%.

TABLE 1
Percent Differences in Side-By-Side Bids by
Contractor for Roof Replacements, 2012–2015.

Contractor Identity	Lowest Bid Difference	Highest Bid Difference	Average Bid Difference
Contractor #1	5.3%	30.1%	12.7%
Contractor #2	1.8%	16.7%	5.5%
Contractor #3	3.4%	33.1%	10.2%
Contractor #4	3.4%	15.4%	11.7%
Contractor #5	0.0%	5.3%	3.2%
Contractor #6	8.9%	42.1%	17.4%
Contractor #7	1.1%	5.7%	3.0%
Contractor #8	8.1%	17.7%	13.5%
Contractor #9	1.5%	26.8%	14.7%
Contractor #10	5.7%	12.5%	9.8%
Overall Averages	3.9%	20.5%	10.2%

Source: Public School Construction Program.

Differences in side-by-side bids may be due to the payment of prevailing wage and benefit rates when contractors plan to use the same workers and production methods on a project. The substitution of skilled for unskilled labor and capital equipment for all grades of labor that typically accompanies wage increases in the construction industry requires time or the entry of contractors with varying skilled workforces and capital intensities (Blankenau and Cassou 2011; Balistreri et al. 2003). The resulting changes in labor productivity and utilization may mitigate some of the cost effect of higher wage rates. However, in the side-by-side bid format, the contactor may face inflexibilities that prevent substitutions with increased wage rates passing directly through to bid costs. This may explain some of the difference in side-by-side bids, but the disparities reported in Table 1 are too large to entirely attribute to labor costs. Many of the "highest bid differences" reported in Table 1 are greater than labor costs for this type of construction activity. Information from the most recent *Economic Census of Construction* indicates that labor costs (wages and benefits) for specialty trade roofing contractors in Maryland are approximately 19.3% of total construction costs(U.S. Bureau of the Census 2012). A bid, like that of Contractor #6 which is 42.1% higher with the payment of prevailing wages is approximately 2.2 times larger than percent labor costs for these types of projects. If the effect of prevailing wages is isolated from other factors that also influence construction costs, the impact of prevailing wages on bids should be fairly uniform from one project and bid to the next. For example, if prevailing wage rates add 10% to the cost of roof replacements, the side-by-side bids should uniformly vary by about 10%, depending on wage differences between counties and over time.

Another possible explanation for varying side-by-side bids is that, while roof replacements are relatively homogenous projects, some may require sheet metal work. Without the payment of prevailing wages, a nonunion contractor would likely have a roofer with suitable experience perform this work with the same rate of pay. But Maryland's prevailing wage regulations, like the federal Davis-Bacon Act and most other state laws, set wage rates for workers performing specific jobs. As a consequence, under the wage policy an employee who splits their time between roofing and sheet metal work must be paid the rates for each job classification accordingly. On average, the total hourly prevailing wage compensation of sheet metal workers is 27.9% higher than the comparable compensation for roofers (State of Maryland, "*Prevailing Wage Information Rates*"). This substantially higher rate may appear to explain some of the bid differences reported in Table 1. However, this implication must be tempered by the fact that labor costs are a low percent of total roofing construction costs. Even if all employees were upgraded to the sheet metal rate, it would affect a relatively small component of total costs and bids. For example, if all roofer labor costs rose by 27.9% to the sheet metal rate and labor costs are 19.3% of total costs, overall costs would increase by about 5.4% (27.9% x 19.3%), assuming that all else is unchanged. The variation in side-by-side bids that cannot be explained by differences in wage rates and the absence of input substitution suggests that factors other than the payment of prevailing wages have an impact on bid differences.

The unbalanced panel of 75 bids by nonunion roofing contractors is used to examine the impact of contractor bid behavior on differences in side-by-side bids by estimating the following one-way fixed effects model:

$$\textit{\% Difference in Bids}_{it} = \beta_0 + \beta_1\,\textit{Contractor}_i + \beta_2\,\textit{\# Bidders}_{it} + \beta_3\,\textit{Bid History}_{it} + \beta_4\,\textit{Bidder Rank}_{it}\,\beta_5\,\textit{Peak Bid Month}_{it} + \beta_6\,\textit{2014 Policy}_{it} + \beta_7\,\textit{Real Midpoint Bid}_{it} + \beta_8\,\textit{County}_{it} + \mu_{it}$$

where *% Difference in Bids* is the difference between the prevailing wage bid and the bid without prevailing wages, divided by the bid omitting prevailing wages (x 100) for roof replacement bids submitted by contractor *i* in time period *t*. *# Bidders* equals the number contractors who submitted a bid for each of the 17 projects. *Bid History* is the accumulated bid experience of each contractor. This information is collected using the longitudinal aspect of the data set where the number of project bids submitted by each contractor is traced from 2012 through 2015.[15] *Bidder Rank* is equal to the order of each bid submitted by the contractors included in the panel. *Peak Bid Month* equals one for bids submitted in March, zero otherwise. *2014 Policy* is a binary variable equal to one for the projects that were completed after the July 1, 2014, prevailing wage policy expansion that lowered the state funding threshold to 25%, zero otherwise. Since this variable captures a time component, year dummy variables are not included for a two-way fixed effects estimate. Since the effects described above may vary with the size of a project, the *Real Midpoint Bid* is added as a control. This variable is the inflation-adjusted midpoint between a contractor's side-by-side bids and allows for the effects of the number of bidders, and bid history, etc. to be measured taking the contractor's perceived value of the project into consideration. *County* is another control variable that takes into consideration regional differences in market and economic conditions. *County* is a dummy variable identifying projects in Carroll, Frederick, and Howard counties with Washington County as the reference category. μ is the error term.

Results

Summary statistics for the variables included in the model are reported in Table 2 (next page). The average difference in prevailing wage bids and bids estimated without the payment of prevailing wages submitted by nonunion roofing contractors is about 10%. Across the ten contractors, this difference was as low as 0.0% and as high as 42%. The number of bidders ranges from two to eight participants per project, with an average of 5.3. The bid history of these contractors is traced longitudinally between 2012 and 2015 and ranges from the first bid to a high of 13 bids with an average of 4.6. It is not possible to determine bid history before 2012, so the measure used here is based on the accumulation of bid experience during the period of the study. The bid ranking of any contractor ranges from the first to the eighth position with an average of about third place. Roofing projects are open to bids in six months of the year (January, February, March, April, August, and December). The peak month for bidding on roof replacement projects is March when 41% of the projects are let and 48% of the bids are placed. One-third (25) of the bids were placed after the policy change in July of 2014 that reduced the state funding threshold to 25% of construction costs. Fifty of the bids were placed under the previous state funding threshold of 50%. The distribution of roof replacements was unevenly distributed with 57% of projects located in Howard County, 21% in Frederick, 16% in Carroll, and 5% in Washington County. The inflation-adjusted midpoint between the bid based on the payment of prevailing wages and the bid omitting the wage requirement is approximately $1.2 million.

Regression results for the fixed effects estimate are reported in Table 3.[16] Because there are *a priori* expectations regarding the effects of the number of bidders, bid history, contractor bid rank, peak bid month, and the 2014 policy change, the coefficients for these variables are evaluated with one-tailed tests. All other coefficients are evaluated with two-tailed tests. Results indicate that the effect of another bidder decreases the gap between bids that are, and are not based on prevailing wage rates by approximately 1.6 percentage points.

Findings also support the notion that as contractors gain experience with side-by-side bidding, the gap between the two bids decreases. The coefficient for bid history reveals that the gap in side-by-side bids decreases by about 1.2 percentage points with each bid experience. The effects of bid competition and bid history are significant at the 0.05 level.

TABLE 2
Summary Statistics of Side-By Side Contactor Bids,
Roof Replacement Projects, Fiscal Years 2012–2015

Variable	**Mean**
% Difference in Bids	9.940 (7.704)
# Bidders	5.293 (1.514)
Bid History	4.640 (2.990)
Bidder Rank	3.107(1.805)
Peak Bid Month (March)	0.480 (0.503)
2014 Policy	0.333 (0.478)
Carroll County	0.160 (0.369)
Frederick County	0.213 (0.412)
Howard County	0.573 (0.498)
Washington County	0.053 (0.226)
Real Midpoint Bid	$1,178,718 (610,602.8)
N	75

Source: Public School Construction Program, State of Maryland.
Standard deviations in parentheses.

TABLE 3
Fixed Effects Regression Results of Side-By Side
Contactor Bids (With and Without Prevailing Wage Rates),
Roof Replacement Projects, Fiscal Year 2012-2015
(Dependent Variable = % Difference in Bids)

Variable	**Coefficient**
# Bidders	–1.558 λλ (0.737)
Bid History	–1.242 λλ (0.532)
Bidder Rank	1.021 λλλ (0.257)
Peak Bid Month (March)	–8.005 λλλ (2.252)
2014 Policy	4.500 λ (2.647)
Carroll County	0.625 (0.369)
Frederick County	13.948 *** (2.181)
Howard County	4.775 ** (2.069)
Real Midpoint Bid	–0.0001 (0.0001)
Constant	18.110** (5.828)
N	75
F	211.24
R^2 (overall)	0.423
F Test, All Individual Effects = 0	4.58

Source: Public School Construction Program, State of Maryland.
Standard errors corrected for heteroskedasticity in parentheses.
λλλSignificant at the 0.01 level (one-tailed test).
λλSignificant at the 0.05 level (one-tailed test).
λSignificant at the 0.10 level (one-tailed test).
***Significant at the 0.01 level (two-tailed test).
**Significant at the 0.05 level (two-tailed test).
*Significant at the 0.10 level (two-tailed test).

Model estimates also support the view that eagerness to win a project affects differences in bids. An increase in bid ranking or place increases the gap by approximately 1 percentage point while side-by-side-bids submitted during the peak month of March are closer by 8 percentage points. Both of these results are significant at the 0.01 level.

Differences in side-by-side bids increased by 4.5 percentage points after the expansion of the prevailing wage policy in 2014. This effect is significant at the 0.10 level. Since the effect of the policy change is measured by comparing bids submitted before and after July 1, 2014, other factors that changed over this time period may also influence the estimated 4.5% increase. One possible influence is the increase in prevailing wage rates over time that would inflate bids if the wage policy applies. However, growth in prevailing wage rates for roofers/waterproofers in the four Maryland counties included in this study was relatively low over the period of the study. Between 2012 and 2015, the prevailing wage and benefit rates for this job classification increased by an average of 3.5%.[17] This increase is substantially lower than the 9.2% increase in the producer price index for roofing contractors over the same period (U.S. Bureau of Labor Statistics, *Producer Price Index by Industry*). These data suggest that prevailing wage growth in Maryland increased proportionately less compared to overall costs for nonresidential roofing contractors. Also, given that labor costs are a low percent of total costs for Maryland roofing contractors, the impact of the increase in prevailing wages on total costs is disproportionately low. If wages increase by 3.5% and labor costs are 19.3% of total costs, the effect of the wage increases is approximately 0.7% (3.5% x 19.3%).[18] Consequently, the change in prevailing wage rates is insufficient to account for the 4.5% increase in side-by-side bids after 2014.

It is also unlikely that the mere expansion of the policy to projects receiving at least 25% in state funding would increase contractor costs and bids. If prevailing wages have a cost impact, it would be measured directly at the level of the project. That is, if a contractor bids on a project that requires prevailing wages and if the contractor expects increased costs as a result, the bid on that project will be higher. The policy change in 2014 would not have an across-the-board impact on project costs and bids. The impact of prevailing wages would still be measured at the project level, regardless of the change in the state funding threshold. Bid costs may increase if the expansion of the policy reduced bid competition. However, the 4.5% increase in side-by-side bids after July 2014 is measured with the level of bid competition held constant.[19] The remaining explanation is that the increase in side-by-side bids is due to the reaction of nonunion contractors who are 'promising' greater saving without the payment of prevailing wages at a time when prevailing wage coverage is expanding.

Holding all other factors constant, differences in side-by-side bids are larger in Frederick and Howard counties compared to Washington County (by about 14 and four percentage points, respectively). While the impacts for these two counties are significant at least the 0.05 level, there is no statistically significant difference in bids between Carroll and Washington counties. The estimate for Real Midpoint Bid is essentially zero in terms of magnitude and statistical significance. This finding indicates that the difference between the two bids does not vary with project size. The results of the F test indicate that the null hypothesis that all coefficients equal zero is rejected at the 0.01 level.[20] The model explains 42% of the total variation in side-by-side bids. The F test implying that individual contractor effects are zero is also rejected at the 0.01 level.[21] This test result indicates that the fixed effects estimate is preferred to an OL estimate that does not control for individual contractor effects.

The results reported in Tables 2 and 3 can be used to illustrate changes in side-by-side bids as the regression equation is solved with a given value of one variable, holding all other variables at their averages. For example, consider changes in the overall average gap in roof replacement bids of 9.9% (as reported in Table 2) when accumulated bid history changes from its average value of 4.64 bids to the maximum number of 13 bids. With the 13th bid the difference between bids based on the payment of prevailing wage and tenders that do not adhere to the wage policy collapses to – 0.4%, holding all else constant. Similarly, if the number of bidders is at its maximum value of eight competitors, bid rank equals first place, and bids are submitted in the peak month of March (with all other variables held at average values), the average gap in side-by-side bids vanishes as the average falls from 9.9% to – 0.6%. While these illustrations do not take into account the confidence intervals of the coefficients or the standard error of the estimate when solving the regression equation, these exercises illustrate the extent to which the difference in bids that are based on the payment of prevailing wages and comparable bids that do not include prevailing wages vary with changes in the bid behavior and outcomes.

Conclusion

The data typically used in studies examining the effect of prevailing wage laws on construction costs allows for the measurement of the policy impact while controlling for other project characteristics that may also be related to costs. The influence of contractor bid behavior has largely been ignored. The side-by-side bid data for Maryland public school tenders provide an opportunity to examine the effects of the level of bid competition, accumulated contractor bid experience, and motivation to win on the relative cost of projects covered by prevailing wage regulations. Results from the fixed effects estimate of an unbalanced panel of roofing contractors indicate that the gap between bids that require and do not require the payment of prevailing wages decreases as the level of bid competition and accumulated contractor bid experience increases. The disparity in side-by-side bids is also influenced by a contractor's eagerness to win a project. Additional analysis illustrates how the average gap between the two bids of 9.9% disappears under particular bid behaviors and outcomes. This evidence from nonunion contractors in Maryland springs from the unique circumstances of this state's prevailing wage policy. Yet, the results provide insights into how contractors, particularly nonunion contractors respond to the requirements of the wage policy.

Endnotes

[1] The Davis-Bacon Act applies to all federally assisted and funded construction with a value in excess of $2,000. See U.S. Department of Labor, "Davis-Bacon and Related Acts." At present, 29 states have prevailing wage laws. Kentucky repealed its prevailing wage law in early 2017, West Virginian repealed in 2016, and Indiana in 2015. For a list of the 18 other states without prevailing wage policies, see U.S. Department of Labor, "Dollar Threshold Amount for Contract Coverage." For an example of a municipal-level prevailing wage standard, see City and County of Denver, "About Prevailing Wage."

[2] As an example, see U. S. Department of Labor, "The Davis-Bacon Act Protecting Wage Equality Since 1931."

[3] See Duncan (2015a) for a comprehensive research review that includes various types of construction.

[4] Change orders and cost overruns may be related to prevailing wage legislation. The single study that has been able to obtain data on cost overruns report that overruns for road construction in Utah tripled in the decade following the 1981 repeal of prevailing wage requirements in this state. See Philips, Mangum, Weitzman, and Yeagle (1995).

[5] See CanaData as an example.

[6] For a description of this policy, see Duncan, Philips, and Prus (2014).

[7] See Ohio Laws and Rules, "Chapter 4115: Wages and Hours on Public Works."

[8] The single study that has examined this issue examines bid behavior as contractors switch from highway resurfacing projects funded by the federal government (and are covered by the Davis-Bacon Act) to projects that are funded by the State of Colorado that are not covered by a prevailing wage standard. See Duncan (2015b).

[9] Information obtained from the Public School Construction Program personnel.

[10] For a description of Maryland's law, see State of Maryland, "Compliance Frequently Asked Questions."

[11] Based on information provided by personnel from the Public School Construction Program, bidders on public works projects in Maryland know the number and identity of bidders for a project. This information would facilitate agreements between contractors.

[12] The data used in this study span 4 years and are insufficient to identify entrant bidders. Others have examined the bids of new contractors. Li and Philips (2012) find that the bids of entrants are more widely dispersed around the central bid tendency. De Silva, Dunne, and Kosmopoulo (2003) find that entrants bid more aggressively than incumbent firms did.

[13] Previous research indicates that bids are higher when a contractors' productive capacity is obligated to previously awarded projects. See Jofre-Bonet and Pesendorfer (2003).

[14] Information from Roofers Local 30, United Union of Roofers, Waterproofers, and Allied Workers of Philadelphia, Pennsylvania, was used to identify roofing contractors who are signatories to collective bargaining agreements. The single union roofing contractor included in the master data file bid on only one project over the time period. According to information provided by the Fair Contracting Foundation, union contractors are hesitant to bid on projects requesting side-by-side bids due to the uncertainty regarding the outcome and whether prevailing wages will be paid or not.

[15] There were a few occasions when projects shared the same bid date. When this is the case, the measure of bid history is the same for both projects.

[16] Standard errors reported in Table 3 are corrected for heteroskedasticity.

[17] In Carroll and Howard Counties, the total prevailing rate (wages and benefits) increased by 2.1% between 2012 and 2015. The corresponding percentage change was 2.4% in Washington County and 7.5% in Frederick County. Data were obtained from State of Maryland, "Prevailing Wage Information Rates."

[18] This method of estimating the increase in total costs due to an increase in prevailing wage rates is over-simplistic as other factors that change with wages (such as labor productivity and use of capital equipment) are ignored. This method is used here to illustrate that the impact of wage increases on total costs is very low.

[19] When the statistical model is estimated without a control for the number of bidders, the measured effect of the 2014 policy change increases to 6.27% with a computed z-statistic of 3.36. Additionally, the two academic studies that examine the effect of prevailing wage laws on bid competition both fail to find a statistically significant impact. See Duncan (2015a) and Kim, Chang, and Philips (2012).

[20] The critical F statistic is 5.35 at the 0.01 level.

[21] The relevant critical F statistic is 2.72 at the 0.01 level.

References

Atalah, Alan. 2013a. "Comparison of Union and Non-Union Bids on Ohio School Facilities Commission Construction Projects." *International Journal of Economics and Management Engineering*, Vol. 3, no. 1, pp. 29-35.

Atalah, Alan. 2013b. "Impact of Prevailing Wages on the Cost among the Various Construction Trades." *Journal of Civil Engineering and Architecture*, Vol. 7, no. 4, pp. 670-76.

Azari-Rad, Hamid, 2005. "Prevailing Wage Laws and Injury Rates in Construction." In Hamid Azari-Rad, Peter Philips, and Mark Prus, eds., *The Economics of Prevailing Wage Laws*. Aldershot, UK: Ashgate, pp. 169-87.

Azari-Rad, Hamid, Philips, Peter and Prus, Mark. 2002. "Making Hay When It Rains: The Effect of Prevailing Wage Regulations, Scale Economies, Seasonal, Cyclical and Local Business Patterns Have On School Construction Costs." *Journal of Education Finance*, Vol. 27, 997-1012.

Azari-Rad, Hamid, Philips, Peter and Prus, Mark. 2003 "State Prevailing Wage Laws and School Construction Costs." *Industrial Relations*, Vol. 42, no. 3, pp. 445-57.

Balistreri, Edward, McDaniel, Christine and Wong, Eina Vivian. 2003. "An Estimation of U.S. Industry-Level Capital-Labor Substitution Elasticities: Support for Cobb-Douglas." *The North American Journal of Economics and Finance*, Vol. 14, no. 3, 343-56.

Belman, Dale. 2005. "Prevailing Wage Laws, Unions, and Minority Employment in Construction." In Hamid Azari-Rad, Peter Philips, and Mark Prus, eds., *The Economics of Prevailing Wage Laws*. Aldershot, UK: Ashgate, pp. 149-68.

Bilginsoy, Cihan. 2005. "Wage Regulation and Training: The Impact of State Prevailing Wage Laws on Apprenticeship." in Hamid Azari-Rad, Peter Philips and Mark Prus, eds., *The Economics of Prevailing Wage Laws*, Aldershot, UK: Ashgate, pp.149-68.

Bilginsoy, Cihan and Philips, Peter. 2000. "Prevailing Wage Regulations and School Construction Costs: Evidence from British Columbia." *Journal of Education Finance,* Vol. 24, pp. 415-32.

Blankenau, William and Cassou, Steven. 2011. "Industry Differences in the Elasticity of Substitution and Rate of Biased Technological Change between Skilled and Unskilled Labor." *Applied Economics*, Vol. 43, pp. 3129-42. CanaData. http://bit.ly/2k2ss50 [January 23, 2017].

City and County of Denver, "*About Prevailing Wage*." Denver Auditor's Office. http://bit.ly/2k2wXwt [January 23, 2017].

De Silva, Dakshina, Dunne, Timothy, and Kosmopoulou, Georgia. 2003. "An Examination of Entrant and Incumbent Bidding in Road Construction Auctions." *The Journal of Industrial Economics*, Vol. 51, no. 3, pp 295-316.

Dodge Data & Analytics. http://construction.com/. [January 23, 2017].

Duncan, Kevin. 2015a. "The Effect of Federal Davis-Bacon and Disadvantaged Business Enterprise Regulations on Highway Maintenance Costs." *Industrial and Labor Relations Review*. Vol. 68, no. 1, pp. 212-37.

Duncan, Kevin. 2015b. "Do Federal Davis-Bacon and Disadvantaged Business Enterprise Regulations Affect Aggressive Bidding? Evidence from Highway Procurement Auction." *Journal of Public Procurement*, Vol. 15, no. 3, pp. 291-316.

Duncan, Kevin, Lantsberg, Alex, and Manzo Frank. 2015. "The Cost of Repealing Michigan's Prevailing Wage Policy: Impacts on Total Construction Costs and Economic Activity." http://bit.ly/2k2kdWQ [January 23, 2017].

Duncan, Kevin, Philips, Peter, and Prus, Mark. 2006. "Prevailing Wage Legislation and Public School Construction Efficiency: A Stochastic Frontier Approach." *Construction Management and Economics*, Vol. 24 (June), pp. 625-34.

Duncan, Kevin, Philips, Peter, and Prus, Mark. 2009. "The Effects of Prevailing Wage Regulations on Construction Efficiency in British Columbia." *International Journal of Construction Education and Research*, Vol. 5, no.1, pp. 63-78.

Duncan, Kevin, Philips, Peter, and Prus, Mark. 2012. "Using Stochastic Frontier Regression to Estimate the Construction Cost Efficiency of Prevailing Wage Laws." *Engineering, Construction and Architectural Management*, Vol. 19, no. 3, pp 320-34.

Duncan, Kevin, Philips, Peter, and Prus, Mark. 2014. "Prevailing Wage Regulations and School Construction Costs: Cumulative Evidence from British Columbia." *Industrial Relations*, Vol. 53, no. 4, October, pp. 593-616.

Jofre-Bonet, Mireia and Pesendorfer, Martin. 2003. "Estimation of a Dynamic Auction Game." *Econometrica*, Vol. 71, no. 5, pp. 1443-89.

Kessler, Daniel and Katz, Lawrence. 2001. "Prevailing Wage Laws and Construction Labor Markets." *Industrial and Labor Relations Review*, Vol. 54, no. 2, pp. 259-74.

Kim, JaeWhan, Chang, Kuo-Liang, and Philips, Peter. 2012. "The Effect of Prevailing Wage Regulations on Contractor Bid Participation and Behavior: A Comparison of Palo Alto, California with Four Nearby Prevailing Wage Municipalities." *Industrial Relations*, Vol. 51, no. 4 (October), pp. 874-91.

Legislative Service Commission. 2002. "S.B. 102 Report. The Effects of the Exemption of School Construction Projects from Ohio's Prevailing Wage Law." http://bit.ly/2k2e5xU [January 23, 2017].

Li, Sheng and Philips, Peter. 2012. Construction Procurement Auctions: Do Entrant Bidders Employ More Aggressive Strategies than Incumbent Bidders?" *Review of Industrial Organization*, Vol. 40, no. 3, pp. 191-205.

Maryland General Assembly. 2016. "Prevailing Wage Law – Applicability to the University of Maryland System, Morgan State University, and St. Mary's College of Maryland." Fiscal and Policy Note, First Reader, Department of Legislative Services. http://bit.ly/2k2IkVg [January 23, 2017].

Ohio Laws and Rules. "Chapter 4115: Wages and Hours on Public Works" http://codes.ohio.gov/orc/4115 [January 23, 2017].

Philips, Peter. 2014. "Kentucky's Prevailing Wage Law: An Economic Impact Analysis." http://bit.ly/2k3f94x [January 23, 2017].

Philips, Peter, Mangum, Garth, Waitzman, Norm, and Yeagle, Anne. 1995. "Losing Ground: Lessons from the Repeal of Nine 'Little Davis-Bacon' Acts." http://bit.ly/2k3nHsh [January 23, 2017].

Public School Construction Program. 2015. "The Cost of School Construction: A Comparison of the Monarch Global Academy and Conventional School Facilities." October 28. http://bit.ly/2k41yK8 [January 23, 2017].

Prus, Mark. 1999. "Prevailing Wage Laws and School Costs: An Analysis for Public School Construction in Maryland and the Mid Atlantic States." http://bit.ly/2k3Hfwv [January 23, 2017].

State of Maryland. "Compliance Frequently Asked Questions (FAQs)—Prevailing Wage for State Funded Construction Contracts." Department of Labor, Licensing, and Regulation. http://bit.ly/2k4HJlV [January 23, 2017].

State of Maryland. "Prevailing Wage – Division of Labor and Industry." http://bit.ly/2k3jvZu [January 23, 2017].

State of Maryland. "Prevailing Wage Information Rates." http://bit.ly/2k2IZpO [January 23, 2017].

U.S. Bureau of the Census. 2012. Economic Census of Construction, Construction: Geographic Area Series: Detailed Statistics for Establishments. http://bit.ly/2k2wudY [January 23, 2017].

U.S. Bureau of Labor Statistics. "Producer Price Index by Industry: Roofing Contractors, Nonresidential Building Work." http://bit.ly/2k0YT3Y [January 23, 2017].

U.S. Department of Labor. "Davis-Bacon and Related Acts." http://bit.ly/2k5yYZ9 [January 23, 2017].

U.S. Department of Labor. "Dollar Threshold Amount for Contract Coverage." http://bit.ly/2k3qq4U [January 23, 2017].

U.S. Department of Labor. "The Davis-Bacon Act: Protecting Wage Equality Since 1931."

Vincent, Jeffrey and Monkkonen, Paavo. 2010. "The Impact of State Regulations on the Cost of Public School Construction." *Journal of Education Finance*, Vol. 35, no. 4 (Spring), pp. 313-30.

Waddoups, C. Jeffrey. 2005. "Health Care Subsidies in Construction: Does the Public Sector Subsidize Low Wage Contractors?" in Azari-Rad, Hamid, Peter Phillips, and Mark Prus, eds. *The Economics of Prevailing Wage Laws,* Ashgate Publishers, pp. 205-24.

XII. LERA Annual Reports

LERA Executive Board Meeting Minutes

10:30 a.m., January 8, 2017
Hyatt Regency, Chicago, IL

In attendance:

- ***Officers:*** *Janice Bellace (President), Bonnie Castrey (Past President), Harry ~~Katz~~ (President-Elect); Craig Olson (Secretary-Treasurer)*
- ***Board Members:*** *Annette Bernhardt, John Budd, Jonathan Donehower, Eric Fidoten, Owen Herrnstadt, Patrice Mareschal, Dan Marschall, Jeff Wheeler*
- ***Other guest:*** *Susan Houseman, LERA Winter Meeting Program Committee Co-Chair*
- ***LERA Staff:*** *Emily Smith, Bernadette Tiemann*

Call to order. The meeting was called to order at 10:28 a.m. by Janice Bellace, President.

Approval of the Minutes

The minutes of the December 2017 LERA Executive Board meeting were reviewed and approved by the Board.

President's Report

ASSA Space Allocation Meeting. LERA attended (Susan Houseman, LERA Program Committee Co-Chair, Janice Bellace, LERA President; and staff Emily Smith and Bernadette Tiemann) the ASSA Space Allocation meeting held Sunday morning, January 8, 2017 in Chicago, and learned that LERA has been removed from their attendance watch list because our attendance has increased dramatically in recent years.

As background, LERA has been conducting sessions at the ASSA meeting for 69 years (since our inception), and LERA is one of the founding ASSA organizations. Up until a few years ago, the ASSA shared revenue with this small group of founding ASSA organizations (about 7 of them), but in 2011, the ASSA decided they no longer wished to continue sharing revenue with this group.

Previously, ASSA had shared revenue with this small group because these founding organizations spent time and resources marketing the event and all contributed to increased registration. Because of their shift in policy in sharing revenue with LERA and the benefits of greater freedom in an independent meeting, LERA decided to continue its sessions at ASSA, but to move special events and the "annual meeting" to the June time-frame, replacing the LERA National Policy Forum. This change in meeting architecture was the result of a review conducted by a special meetings architecture committee which brought their recommendations to the board in 2011, and the new meeting architecture begin in 2012. In January of 2012, LERA met with the ASSA, and conducted our 27 sessions as usual, but moved all its special events to the new independent Annual Meeting, which was held May 2013 in St. Louis, MO.

In 2012 and 2013, LERA's session attendance dropped at ASSA, and ASSA subsequently reduced our sessions from 27 to 18. The Program Committee, then headed by Eileen Appelbaum felt this was an appropriate fit for LERA, and that 27 was unsustainable number of sessions. The program committee then began working towards the goal of increasing attendance at the LERA@ASSA in 2016 and 2017 and our session attendance improved, and the ASSA has now removed LERA from its watch list.

Webcasting. Bernadette Tiemann summarized initiative of webcasting preconference developmental workshops and it is decided to pursue software from University of Illinois and trying in Anaheim.

Expanding Onsite Visibility, Janice Bellace, President discussed the possibility of having an internship for building an app for smartphone use. Eric Fidoten commented that the value of an organization is human capital. Let's value webcasting and expand webcasting over building an online application for LERA for now.

- Training for unions to webcast from LERA pre-conference professional development workshops
- How do we attract more union officials to LERA programming?
- Favorable towards proposal- explore e-options, look into a proposal by Anaheim (punch-free domain, ask Eric Fidoten.)

Implementation Committee's Report

Bonnie Castrey reported that the Implementation Committee recommends:

1. To set Apprenticeship student dues at the same level as Student Member dues. This motion passed unanimously.
2. That LERA will send a list of potential members, who do not belong to a local chapter, to local chapters, twice annually. The board discussed the implementation of this and suggests that we - prepare a draft letter about joining to distribute to the committee; opting to test out a soft-approach.
3. To establish an evolving understanding of what our new Regional Vice Presidents will set out to accomplished; motion passed. *Discussion:* Regional VP's are created to
    a. Enhance communication between national and chapters
    b. Organize regionally
    c. Serve as a member of this board and of the National Chapter Advisory Council.
4. Implementation Committee and Regional VP's will begin to create structures and duties as the process evolves. Jeff Wheeler suggested we simplify roles and objectives, and that we specify the meetings that these individuals are expected to attend annually.
5. Emily Smith and Joel Cutcher-Gershenfeld will put together a survey for chapters and help connect chapters and help connect chapters with list serves, webcasting etc.
6. Motion initial start-up phase – strive for achieving in sectoral diversity, but primary goal is to enhance communication and we will be flexible to achieve that.
7. First term for Regional VP's will be 3 years as well as thereafter
8. Regional VP's will need to be elected nationally; we will invite them to attend our June meetings so members can meet the candidates.
9. In our bylaws, we use the nomenclature "Affiliates" instead of "Affiliate Members". National members are members of LERA, Chapter members are members of a LERA chapter, and chapter members are also affiliates of the national organization.
10. Committee recommends working with the secretary/treasurer and Emily Smith to determine if a dues increase for chapters is recommended instead of having individual "head tax" for affiliates; motion passed unanimously and some of the discussion was:
    - Chapters are currently reviewing their own dues structures
    - Let's begin with a comprehensive and inclusive pricing policy (Chapter dues covers the cost of affiliates) and then invite these people to participate in activities that raise revenue
    - UCIRHRP Chairs need to be invited to sit on the LERA Executive Board each year
    - 2017 LERA President Elect will Chair the LERA 70th Annual Meeting Program Committee and that structure will continue going forward, although the President will have final say and input.

Financial and Membership Report

Craig Olson, LERA Secretary-Treasurer, reported that overall our budget is in the black by around $23,000, our first surplus in about ten years. This is due to having only one full-time staff member at LERA and increasing revenue associated with the LERA Annual Meeting, and also because we saved money in a few areas.

Sponsorships will be an important part of the LERA 2017 budget, at $30,000 factored in, but we do not know yet where this sponsorship will come from, and we hope that Kaiser Permanente will again sponsor LERA. Owen Herrnstadt will talk to LERA staff about contacting Ford again about sponsoring LERA in 2017.

A second full-time LERA staff member is factored into the proposed 2017 budget; this is an important measure as overtime can only carry the organization so far.

In terms of expenses, LERA saved some money in 2016. LERA's publication costs were only two-thirds of projections with savings reported for both the LERA Research Volume and the *Perspectives on Work* magazine. Some of the cost savings was achieved by carefully reviewing printing bids and estimates. LERA also decreased the number of pages in LERA Perspectives on Work magazine Volume 20 a bit which reduced production costs and allowed us to bind the magazine more cheaply. At the same time, the production team introduced a new layout on the cover, and included award winners, which we think is an attractive and sustainable format. LERA could entertain the idea of giving members an option to opt out of print versions because they can view them online in the future, as an additional cost-savings measure to the organization.

The 2017 budget is structured to break-even, and to achieve this, increases in meeting registration and membership dues are recommended, to begin in January 2017. It is further recommended that a focus group review a comprehensive dues plan for 2018 and beyond. Harry Katz motioned to accept the proposed dues increase be approved and authorize a study group to review dues models (including a comprehensive evaluation beginning with how other comparable association's price membership, sliding dues scales based on income or other structures, increasing dues levels, charging extra for optional components such as Interest Sections, Industry Councils, printed publications, etc., and additional review items such as determining if we are the primary or secondary professional organization for LERA members, etc.); motion passed unanimously.

Other outcomes from this discussion were: Annette Bernhardt and Eric Fidoten should be added to the Development Committee, a menu of sponsor-able activities should be produced, and a 75th Capital Fund campaign should be reviewed by the Development Committee.

LERA Winter Meeting Program Committee Report

Susan Houseman reported that after the shift in meeting architecture in 2012 (to move LERA Annual Meeting in May/June, but keep the sessions at the ASSA meeting in January) attendance has dropped at the ASSA meeting. The program committee's strategy they began was to mainly recruit sessions, specifically sessions of interest to labor economists. This last year, of the 18 sessions that LERA produced at the LERA Winter Meeting, 12 of them were recruited by the program committee members, 6 of them were submitted (and only 3 of those accepted) and the remaining sessions were made of individually submitted papers.

The 2018 Strategy will focus more on reaching out to more colleagues for submitted sessions (over specifically recruited sessions), and the program committee will firmly require LERA membership of at least one author on any paper that will be presented on the program. The program committee is considering running an ad in the Journal of Economics to advertise LERA and our open CFP for 2018, and possibly the ILR Review or the Journal of Labor Economics. They will also make the CFP widely available and send it to the LERA Board members.

LERA Annual Meeting Program Committee Report

Janice Bellace, LERA President and Program Committee Chair reported that supply chain technology is going to be a topic of discussion, as will outsourcing/public sector/Davis Bacon/unpaid interns at Disney. Janice Bellace and Jeff Wheeler will pursue both a session and a sponsorship in reference to Disneyland. To make ends meet at the Hilton Anaheim, where our contract for BEO is a minimum of $50,000, we have two provisions.

The first is a recommendation to increase early bird regular member full conference registration from $295 to $335; that motion seconded by Jonathan Donehower and passed unanimously. The second is a proposal by Kaiser Permanente to conduct a small meeting of about 30 KP staff members to contribute about 10 to 12,000 towards LERA's BEO minimum. This suggestion has also been approved by the LERA Board.

The LERA Board and program committee will continue to think about how to raise money for speaker registration as this topic is brought up each year by various chairs and session organizers who would like to include people in the program who may not be able to afford to pay the full registration price. In terms of marketing the event, we need to make it clear to people to attend the event that it is very important that they get the message that LERA needs them to stay at the LERA hotel, or we may not meet our hotel room block as it is very high this year.

Upcoming LERA Deadlines and Meetings

- LERA 69th Annual Meeting – Thu.-Sun., Jun. 1-4, 2017 (Pre-Conf.: May 31) at the Anaheim Hilton
- Awards nominations due – Sun., Jan. 15, 2017
- Chapter award nominations due – Wed., Feb. 15, 2017
- Winter meeting session proposals due – Thu., Mar. 9, 2017
- Executive Board Meeting – Fri., Jun. 2, 2017, over lunch, at the Anaheim Hilton

President Bellace adjourned the meeting at 1:50 p.m.

LERA Executive Board Meeting Minutes

Noon to 2 p.m., June 2, 2017
Malibu Room, Hilton Anaheim

In attendance:

- ***Officers:*** *Janice Bellace, President, Harry Katz, President Elect; Bonnie Castrey, Past President; Kris Rondeau, President Elect-Elect and current board member; Bill Canak, NCAC Chair; Craig Olson, Secretary-Treasurer; and Ariel Avgar, Editor-in-Chief*
- ***Board Members:*** *John Budd, Dennis Dabney, Jonathan Donehower, Rebecca Givan, Owen Herrnstadt, Charles Jeszeck, Dan Marschall, Saul Rubinstein, Jeff Wheeler*
- ***Committee Chairs:*** *Joel Cutcher-Gershenfeld, IC/IS Coordinating Committee Co-Chair; Marlene Heyser and Jim Pruitt, Development Committee Co-Chairs; David Lewin, Strategic Thinking; and Jeff Keefe, Membership*
- ***Special Guests:*** *Incoming Board Members Michele Hoyman, Joan Husted, and Sheila Mayberry*
- ***LERA Staff:*** *Emily Smith and Bernadette Tiemann*

Unable to attend: *Annette Bernhardt, James Hayton, Patrice Mareschal, Beth Schindler*

Call to order. The meeting was called to order at 11:55 a.m. by Janice Bellace, President.

Approval of the Minutes

The minutes of the January 2017 LERA Executive Board Meeting and the June 2016 General Membership Meeting Minutes were reviewed and unanimously approved by the Board.

Committee Reports

Financial Report—reported by Craig Olson, LERA Secretary-Treasurer. The financial report was reviewed mid-year 2017 figures, and accepted by the board. Mid-year income is on track with projections, although sponsorships are still unaccounted by and if not procured, will be a large impact to our bottom line. Dennis Dabney pledged support on behalf of Kaiser Permanente in the amount of $20,000 sponsorship in 2017. Saul Rubinstein suggested analyzing whether a switch in our financial calendar to a fiscal year would better suit LERA's needs and meeting structure.

Strategic Thinking Committee/Implementation Committee Reports—reported on by David Lewin and Bonnie Castrey, Chairs. The LERA Strategic Thinking Committee reported on a comprehensive set of bylaws revisions that were voted on in the LERA General Membership meeting in June 2016 (that subsequently passed). David explained the architecture of the new meeting structure in separating from ASSA and the implementation of those changes. Additionally, he explained the connection between local chapter members and national members. The bylaws revisions were approved in Minneapolis; changes have been slow and steady in implementation. We are now welcoming our three new RVPs to the board who have just been elected. The slowest implementation is in obtaining chapter member lists. David also commented that the NAA is in favor of continuing to collaborate in meeting speakers/sessions with LERA. Saul Rubinstein suggested considering integrations with other organizations, specifically the Canadian LERA. Bonnie Castrey commented that we had inquired, but that Canadian LERA said they weren't interested for several years.

Industry Council/Interest Section Coordinating Committee Report—reported on by Joel Cutcher-Gershenfeld, Chair. Originally 8 Industry Councils were established. Three of these have not demonstrated activity of late (auto, public sector, and airlines). Three may need to be organized (sports, arts & entertainment; IT; hospitality and services). Joel suggested that future programs reflect any association with Industry Councils or Interest Section. Rebecca Givan suggested using an "app" to help create "tracks". There has been growth in the Industry Council sites to include updates and dialog. The LERA certification initiative also continues to be of interest.

Development Committee Report—reported on by Marlene Heyser and Jim Pruitt, Co-Chairs. The committee motioned to liquidate $4,000 from the Dunlop and Gershenfeld funds to be used toward those expenses in the LERA budget; the motion passed unanimously. It was suggested that board members help with fundraising by bringing in organizational memberships and personal contributions. Consider approaching unions for organizational memberships, and touching base with expired organizational members. Jim Pruitt requested the expired organizational member list. Janice Bellace suggested starting a 70th Anniversary fund. Marlene Heyser briefly updated the board on the state of the potential project with BlueCross BlueShield NLO. The project has not had forward momentum to date.

Membership Committee Report—reported on by Jeff Keefe, Chair. LERA has experienced a 9-10% growth in membership since Emily took over, but there is no constant growth strategy. Jeff Keefe suggests we need out and build our managerial memberships; that they have decreased over the years, and that w consider, what are we doing programmatically. Do we have to have content that is valuable to them a members? He also suggested that we move towards a LERA journal to help build young members (non-tenure track academics), or to possibly adopt a current journal that could benefit from the backing of LERA.

Editorial Committee Report—reported on by Ariel Avgar, LERA Editor-in-Chief. ILR review competition has launched in 2017. The ILR Special LERA Best Papers volume has had two accepted submissions this year. The research volume continues to speed up; editorial committee has moved towards a model of two working RV's so that we are ahead of the production timeline. The board discussed the proposal 2019 RV; it was unanimously approved move forward with the discussed changes: scholars from varying universities, and an international focus. Ariel suggested an award for best book(s) in the field. The editorial committee will work on the logistics and make a proposal to the board.

National Chapter Advisory Council Report—reported by Bill Canak, NCAC Chair. Bill shared the 2017 LERA Chapter Awards results. Additionally, he introduced the three new RVP's that have now taken their place at the board. One chapter has moved to inactive status, and one close by chapter has been revived. We continue to work out the implementation of providing benefits to LERA chapter affiliates.

Annual Meeting Program Committee Report—reported by Kris Rondeau and Harry Katz, Co-Chairs. The local Baltimore planning team shared a progress update, and the Hilton Baltimore contract was reviewed. Two planning meetings are scheduled for June 2017 and two additional planning meetings will take place via conference call. John Budd suggested a slow increase in registration instead of another large jump in the future.

LERA@ASSA Meeting Program Committee Report—reported by Janice Bellace, President. In an effort to make the most of the LERA@ASSA meeting, the program committee has recently begun to require and enforce a requirement of LERA membership to present a paper at the meeting. Janice Bellace also noted that the new branding is under way. Rebecca Givan asked that the Call for Proposals be upfront about the draw for economists.

Upcoming Dates

The LERA Executive Board will again meet in Philadelphia in conjunction with the LERA@ASSA Meeting on Sunday, January 7, 2018, and then in Baltimore, MD in conjunction with the LERA 70th Annual Meeting, on Friday, June 15, 2018.

The meeting was adjourned at 1:55 p.m. by Janice Bellace.

LERA General Membership Meeting and Awards Ceremony

5:30 p.m., June 3, 2017
California Ballroom C, Hilton Anaheim

Call to order. The meeting was called to order at 5:34 p.m. by Janice Bellace, President

Committee Reports

Nominating Committee Report. Janice Bellace reported the outcome of the most recent LERA Election and welcomed the newest board members from the 2017 election.

Finance and Membership Report. Craig Olson reported that we have made progress again to recover from from past years' deficits. Several sponsors have renewed their commitment to the organization including including Kaiser Permanente and Ford/UAW, and our cohort of University programs. Kaiser Permanente helped to underwrite the LERA 69th Annual Meeting, holding a meeting in conjunction with our own, and in doing so, helped LERA pay for an otherwise expensive meeting contract at the Hilton Anaheim. Dues for 2017 will be held at the current levels. With the new Chapter affiliate initiative, the association is hoping to bridge the communication gap between chapter members and the national organization.

Development and Contributions Committee Report. Marlene Heyser reported that LERA received major sponsorships from both Kaiser Permanente and Ford/UAW this year, and that the development committee will begin a new funding initiative to raise funds this year and honor-a-member at the same time. Additionally, there will be an effort to educate LERA members to how to leave a gifts from their estates.

Editorial Committee Report. Ariel Avgar, Editor in Chief reports that the 2017 Research Volume will be a volume edited by Johanna Westar, "The Contradictions of Pension Fund Capitalism". The Committee is currently preparing concepts for the 2018 and 2019 volumes, including Janice Fine's proposal for a volume on "No One Size Fits All: Worker Organization, Policy and Movement In a New Economic Age", which had initially been tapped as the 2017 edition, but production changes pushed it into 2018. Avgar reported that another task of the Editorial Committee is to pre-review submitted papers for the Best LERA Papers to be published in an *ILR Review* issue. The Committee thanked all who submitted papers for Competitive Papers and for the many paper sessions in the meeting.

Program Committee Report. Harry Katz, President Elect and Program Chair for the LERA 2017 Annual Meeting in Anaheim CA, on June 1-4 at the Hilton Anaheim Hotel shared that the meeting theme is "Jobs, Opportunity and Equity in the New World of Work" and the call for proposals has been issued. Deadline is November 15, 2016.

NCAC Chapter Report and Star Award. Bill Canak, Chair of the National Chapter Advisory Committee, reported on health of LERA Chapters, which are in good shape. A new startup of a Minnesota Chapter could be in the works, and LERA has it's first new student chapter in New Jersey (RU LERA). LERA has now elected our first three Regional Vice Presidents who will work to organize LERA Chapters regionally and provide additional resources to the chapters in their regions, as well as provide a conduit of information between the national organization and the local LERA Chapters. Each Regional Vice President (one from Mid region, one from East region, and one from West region) has a vote on the LERA board and is charged to represent LERA chapter interests. The names of the people who ran for this position were brought forth from the chapter level of the organization, under the direction of the NCAC, and were openly elected. He presented a 2017 Chapter Star Award to both TERRA and Atlanta LERA. Additionally, Long Island LERA received a LERA Outstanding Chapter Award, and LERA Chapter Merit awards were given to Mid-Michigan

LERA, Northwest LERA, Orange County LERA, Oregon LERA, LERA Gateway, RU LERA, Alabama LERA, Southwestern PA LERA, and PA Laurel Highlands LERA.

Implementation Committee Report. Bonnie Castrey, Past President and Chair explained a new initiative from the committee and Board is meant to involve chapter members in the national LERA and requiring national members to belong to a chapter if a chapter is located their area. A new "affiliate" status would be given to all chapter members allowing them access to all LERA publications, news, job board, and other services available from the LERA website. Additionally, the Board will be expanded to include three new regional vice presidents from the chapters, alternating regions and perspectives. The motion was made.

2017 Awards Ceremony

Media Award: David Lewin of the Media Award Committee presented the award to Peter Waldman, Bloomberg and Natalie Kitroeff, Los Angeles Times.

Best Dissertation Award Committee. Jeff Keefe of the Thomas A. Kochan and Steven R. Sleigh Best Dissertation Award Committee awarded the Best Dissertation Award to: Bradley Weinberg, Cornell. Two honorable mentions were also awarded to Jae Eun Lee, Ithaca College, and Akasemi Newsome, UC, Berkeley.

LERA Awards Committee. Co-Chairs Doug Kruse and Tia Denenberg presented the 2017 slate of awards. The John T. Dunlop Outstanding Scholar Award was given to Victor Tan Chen, Virginia Commonwealth University for national research issues and the winner for international contributions was Dionne Pohler, Univ. of Toronto. The Outstanding Practitioner Awardee was Sheldon Freidman, AFL-CIO The Susan C. Eaton Outstanding Scholar-Practitioner Award went to Erica Groshen, Former Commissioner or BLS.Academic Fellows for 2017 are William Holley Jr., Michael Reich, Adrienne Eaton. Practitioner Fellows are Mark Cousens, Bill Dirksen, and Mary Ellen Shea.

James Scoville Best International Paper Committee. Committee chair, John Budd presented the 2017 Best International Paper Award to co-winners: Peter Sheldon, Raoul Nacamulli, Francesco Paoletti, and David E. Morgan.

Susan C. Eaton Grant Award. Committee Chair, Steve Sleigh presented the 2017 Susan C. Eaton Grant Award to Kyle Albert, Harvard University.

LERA Chapter Star Award. NCAC chair, Bonnie Castrey presented the 2017 LERA Chapter Star award to TERRA, Bill Canak accepted the award, and LERA-Atlanta, award was announce, but physical award will be presented at Georgia Labor-Management Conference later this June.

New and Other Business

The next LERA Annual Meeting will be held June 14-17, 2018 in Baltimore, MD, in the Hilton Baltimore Hotel. After asking for any new or other Business items and hearing none, President Janice Bellace thanked the Executive Board and all present and handed over the gavel to Harry Katz, the new LERA President. President Katz adjourned the meeting at 7:25 p.m.

Labor and Employment Relations Association

Champaign, Illinois

Financial Statements

For the Years Ended December 31, 2016 and 2015

Feller & Kuester CPAs LLP
Certified Public Accountants
806 Parkland Court, Suite 1
Champaign, IL 61821
217-351-3192

TABLE OF CONTENTS

Feller & Kuester CPAs LLP

Tax - Audit - Bookkeeping
806 Parkland Court, Suite #1 - Champaign, Illinois 61821
Phone - (217) 351-3192 Fax - (217) 351-4135 Email - neal@fellerkuester.com

INDEPENDENT AUDITOR'S REPORT

To the Board of Directors of
Labor and Employment Relations Association
Champaign, Illinois

We have audited the accompanying financial statements of the Labor and Employment Relations Association (a nonprofit organization) which comprise the statements of financial position as of December 31, 2016 and 2015, and the related statements of activities and cash flows for the years then ended, and the related notes to the financial statements.

Management's Responsibility for the Financial Statements
Management is responsible for the preparation and fair presentation of these financial statements in accordance with accounting principles generally accepted in the United States of America; this includes the design, implementation, and maintenance of internal control relevant to the preparation and fair presentation of financial statements that are free from material misstatement, whether due to fraud or error.

Auditor's Responsibility
Our responsibility is to express an opinion on these financial statements based on our audits. We conducted our audits in accordance with auditing standards generally accepted in the United States of America. Those standards require that we plan and perform the audit to obtain reasonable assurance about whether the financial statements are free from material misstatement.

An audit involves performing procedures to obtain audit evidence about the amounts and disclosures in the financial statements. The procedures selected depend on the auditor's judgment, including the assessment of the risks of material misstatement of the financial statements, whether due to fraud or error. In making those risk assessments, the auditor considers internal control relevant to the entity's preparation and fair presentation of the financial statements in order to design audit procedures that are appropriate in the circumstances, but not for the purposes of expressing an opinion on the effectiveness of the entity's internal control. Accordingly, we express no such opinion. An audit also includes evaluating the appropriateness of accounting policies used and the reasonableness of significant accounting estimates made by management, as well as evaluating the overall presentation of the financial statements.

We believe that the audit evidence we have obtained is sufficient and appropriate to provide a basis for our audit opinion.

Opinion

In our opinion, the financial statements referred to above present fairly, in all material respects, the financial position of the Labor and Employment Relations Association as of December 31, 2016 and 2015, and the changes in its net assets and its cash flows for the years then ended in accordance with the accounting principles generally accepted in the United States of America.

Other Matter

Our audit was conducted for the purpose of forming an opinion on the financial statements as a whole. The schedules of functional expenses on pages 16 and 17 are presented for purposes of additional analysis and are not a required part of the financial statements. Such information is the responsibility of management and was derived from and relates directly to the underlying accounting and other records used to prepare the financial statements. The information has been subjected to the auditing procedures applied in the audit of the financial statements and certain additional procedures, including comparing and reconciling such information directly to the underlying accounting and other records used to prepare the financial statements or to the financial statements themselves, and other additional procedures in accordance with auditing standards generally accepted in the United States of America. In our opinion, the information is fairly stated in all material respects in relation to the financial statements as a whole.

Feller & Kuester CPAs LLP

Feller & Kuester CPAs LLP
Champaign, Illinois

April 28, 2017

LABOR AND EMPLOYMENT RELATIONS ASSOCIATION
STATEMENTS OF FINANCIAL POSITION
DECEMBER 31, 2016 AND 2015

	2016	2015
ASSETS		
Current Assets		
Cash and Cash Equivalents	$ 167,732	$ 136,263
Investments	23,796	23,145
Accounts Receivable, Net	-	911
Prepaid Expenses	10,972	2,296
Inventory	6,622	6,729
Accrued Interest and Royalties	-	2,617
Total Current Assets	209,122	171,961
Property and Equipment, Net	276	848
TOTAL ASSETS	$ 209,398	$ 172,809
LIABILITIES AND NET ASSETS		
Current Liabilities		
Accounts Payable	$ 15,421	$ 575
Funds Held for Other Organizations	6,605	5,805
Accrued Liabilities	20,424	15,557
Dues Collected in Advance	61,413	54,742
Subscriptions Collected in Advance	13,584	10,484
Deferred Chapter Dues	7,800	6,300
Other Deferred Revenue	1,500	-
Total Current Liabilities	126,747	93,463
Net Assets		
Temporarily Restricted		
Susan C. Eaton Scholar-Practitioner Memorial Fund	32,413	32,040
Kochan-Sleigh Best Dissertation Award Fund	41,941	43,090
Jim Scoville Award Fund	5,251	6,455
John T. Dunlop Public Policy Fund	20,657	19,809
Gladys and Water Gershenfeld Publication Fund	24,550	23,298
Total Temporarily Restricted	124,812	124,692
Unrestricted Net Assets	(42,161)	(45,346)
Total Net Assets	82,651	79,346
TOTAL LIABILITIES AND NET ASSETS	$ 209,398	$ 172,809

See Accompanying Notes.

LABOR AND EMPLOYMENT RELATIONS ASSOCIATION
STATEMENTS OF ACTIVITIES
FOR THE YEARS ENDED DECEMBER 31, 2016 AND 2015

	2016			2015		
	Unrestricted	Temporarily Restricted	Total	Unrestricted	Temporarily Restricted	Total
Support and Revenues						
Membership Dues	$ 115,654	$ -	$ 115,654	$ 118,190	$ -	$ 118,190
Meeting Income	140,661	-	140,661	87,484	-	87,484
Organization Dues & Sponsorships	40,000	-	40,000	49,000	-	49,000
Subscriptions	18,035	-	18,035	18,546	-	18,546
Chapter Fees	9,750	-	9,750	12,760	-	12,760
Member Contributions	13,040	1,570	14,610	11,499	315	11,814
Royalties	4,951	-	4,951	5,772	-	5,772
Ad Income	4,013	-	4,013	1,400	-	1,400
Publications	446	-	446	257	-	257
Mailing List Rental	434	-	434	238	-	238
Administrative Fees	174	-	174	98	-	98
Investment Income	-	651	651	-	289	289
Interest Income	140	323	463	314	177	491
Net Assets Released from Restrictions	2,424	(2,424)	-	1,193	(1,193)	-
Total Support and Revenues	349,722	120	349,842	306,751	(412)	306,339
Expenses						
Program Services						
General	156,097	-	156,097	166,237	-	166,237
Meetings	92,726	-	92,726	89,809	-	89,809
Publications	31,297	-	31,297	41,604	-	41,604
Supporting Services						
Management and General	51,659	-	51,659	50,979	-	50,979
Membership Development	14,758	-	14,758	6,712	-	6,712
Total Expenses	346,537	-	346,537	355,341	-	355,341
Change in Net Assets	3,185	120	3,305	(48,590)	(412)	(49,002)
Net Assets, Beginning of Year	(45,346)	124,692	79,346	3,244	125,104	128,348
Net Assets, End of Year	$ (42,161)	$ 124,812	$ 82,651	$ (45,346)	$ 124,692	$ 79,346

See Accompanying Notes.

LABOR AND EMPLOYMENT RELATIONS ASSOCIATION
STATEMENTS OF CASH FLOWS
FOR THE YEARS ENDED DECEMBER 31, 2016 AND 2015

	2016	2015
CASH FLOWS FROM OPERATING ACTIVITIES		
Change in Net Assets	$ 3,305	$ (49,002)
Adjustments to Reconcile Change in Net Assets to Net Cash Provided by (Used in) Operating Activities		
Depreciation Expense	572	739
Net Unrealized (Gains) Losses on Investments	62	412
(Increase) Decrease in Operating Assets		
Accounts Receivable, Net	911	(661)
Prepaid Expenses	(8,676)	3,960
Inventory	107	(1,192)
Accrued Interest and Royalties	2,617	1,688
Increase (Decrease) in Operating Liabilities		
Accounts Payable	14,846	(7,510)
Funds Held for Other Organizations	800	(1,711)
Accrued Liabilities	4,867	(8,758)
Dues Collected in Advance	6,671	(2,015)
Subscriptions Collected in Advance	3,100	3,206
Deferred Sponsorships	-	(25,000)
Deferred Chapter Dues	1,500	(2,260)
Other Deferred Revenue	1,500	(1,755)
Net Cash Provided by (Used in) Operating Activities	32,182	(89,859)
CASH FLOWS FROM INVESTMENT ACTIVITIES		
Reinvested Interest, Dividends, and Capital Gains	(713)	(701)
Net Cash Provided by (Used in) Investment Activities	(713)	(701)
CASH FLOWS FROM FINANCING ACTIVITIES		
None	-	-
Net Cash Provided by (Used in) Financing Activities	-	-
Net Increase (Decrease) in Cash and Cash Equivalents	31,469	(90,560)
Beginning Cash and Cash Equivalents	136,263	226,823
Ending Cash and Cash Equivalents	$ 167,732	$ 136,263

See Accompanying Notes.

LABOR AND EMPLOYMENT RELATIONS ASSOCIATION
NOTES TO FINANCIAL STATEMENTS
DECEMBER 31, 2016 AND 2015

NOTE 1 – NATURE OF ACTIVITIES AND SIGNIFICANT ACCOUNTING POLICIES

Organization and Nature of Activities

The Labor and Employment Relations Association (the Organization) was founded in 1947 to encourage research in all aspects of the field of labor, employment, and the workplace. It is a nonprofit scholarly association of academic, labor, business, and neutral communities committed to the full discussion and exchange of ideas between and among its broad constituencies through meetings, publications, and its various electronic listservs and websites. The Organization's National Office is located in Champaign, Illinois and serves the association by planning conferences and meetings and publishing the various researches of its members. The main sources of support and revenues for the Organization are contributions, membership dues and subscriptions, and meeting income.

Subsequent Events

Subsequent events have been evaluated through April 28, 2017, which is the date the financial statements were available to be issued.

Accrual Basis of Reporting

The Organization has chosen to report on the accrual basis of accounting. Accordingly, revenue is recognized when earned and expenses are recognized when incurred in conformity with accounting principles generally accepted in the United States of America (GAAP).

Estimates

The preparation of financial statements in conformity with GAAP requires management to make estimates and assumptions that affect certain reported amounts and disclosures. Accordingly, actual results could differ from those estimates.

Cash and Cash Equivalents

For purposes of reporting cash flows, cash and cash equivalents include all cash and highly liquid investments acquired with an original maturity date of three months or less. Since the penalties of converting certificate of deposits to cash is insignificant, all certificate of deposits have been included with cash and cash equivalents. As of December 31, 2016 and 2015, $101,016 and $101,547 of cash and cash equivalents are temporarily restricted and are subject to donor-imposed stipulations.

Investments

Investments consist of intermediate term bond index funds. All investments are held as temporarily restricted and are subject to donor-imposed stipulations.

Accounts Receivable

Accounts receivable are recorded primarily for outstanding invoices for membership dues and subscriptions. An allowance for doubtful accounts is based on an analysis of expected collection rates determined from experience. The Organization had an allowance for doubtful accounts of $100 at December 31, 2016 and 2015.

Inventory

The Organization's inventory of directories, research volumes, proceedings, and perspective magazines is carried at the lower of cost and market value. Cost is determined on the basis of first in – first out.

Property and Equipment

Property and equipment expenditures in excess of $500 are capitalized at cost. Donated property and equipment are capitalized at estimated cost or fair market value at the time of donation. Depreciation of the assets is computed using the straight-line method over their estimated useful lives. The range of estimated useful lives by type of asset is as follows:

Furniture and Equipment	5-7 years

Net Assets

Net assets of the Organization and changes therein are classified and reported as follows:

Unrestricted Net Assets – Net assets that are not subject to donor-imposed stipulations.

Temporarily Restricted Net Assets – Net assets subject to donor-imposed stipulations that may or will be met, either by actions of the Organization and/or passage of time. When a restriction expires, temporarily restricted net assets are reclassified from restrictions. Contributions received with temporarily restrictions that are met in the same reporting period are reported as unrestricted support and increase unrestricted net assets. The Organization had $124,812 and $124,692 of temporarily restricted net assets as of December 31, 2016 and 2015, respectively.

Permanently Restricted Net Assets – Net assets subject to donor-imposed stipulations that they be maintained permanently by the Organization. Generally, the donors of these assets permit the Organization to use all or part of the income earned on any related investments for general or specific purposes. The Organization had no permanently restricted net assets as of December 31, 2016 and 2015.

Grant Revenue

Grant funds received under contracts that require provision of services by the Organization are recognized as revenue to the extent that expenses have been incurred for the purpose specified by the grantor. The excess of grant monies received over expenses incurred is accounted for as deferred revenue. The excess of grant expenditures over grant monies received is accounted for as grant revenue receivable to the extent that additional grant revenue has been awarded and will be paid.

Grant funds received with only general restrictions and public donations are considered to be available for unrestricted use unless specifically restricted by the donor. When such restrictions exist, the revenue is recorded as temporarily restricted until the restrictions are met.

Contributed Services and Goods

Contributed services are reported as contribution revenue and as assets or expenses only if the services create or enhance a non-financial asset (for example property and equipment) or:

- Would typically need to be purchased by the Organization if the services had not been provided by contribution.
- Require specialized skills.
- Are provided by individuals with those skills (such as accounting, financial, construction, educational, electrical, legal, medical, and other services provided by accountants, investments advisers, contractors, teachers, electricians, lawyers, doctors, and other professional and craftspeople).

For the years ended December 31, 2016 and 2015, the value of contributed services meeting the defined requirements for recognition in the financial statements as outlined above were not material and have not been recorded on the financial statements. The Organization does receive free office space from the University of Illinois. However, the value for the use of this office space has not been reported on the financial statements.

Income Tax Status

The Organization is a nonprofit association that is exempt from federal income tax under Section 501 (c)(3) of the Internal Revenue Code. In addition, the Internal Revenue Service has determined that the Organization is not a private foundation as defined in Section 509(a)(1) and Section 170(b)(1)(A)(vi) of the Code.

The Organization has evaluated its exposure resulting from uncertain income tax position and determined the exposure is not material to the financial statements. In addition, the Organization is not aware of any tax position for which a significant change is reasonably possible within the next 12 months. Therefore, these financial statements do not include a liability for uncertain tax positions. Upon recognition of a liability for an uncertain tax position, the Organization would recognize interest expense and penalties in operating expenses.

The Organization files information tax returns in the U.S. federal jurisdiction and the state of Illinois. Its federal and Illinois information tax returns prior to fiscal year 2013 are closed. The Organization does not have any tax returns currently under examination by either the Internal Revenue Service (IRS) or any U.S. state jurisdiction.

Membership Dues and Advance Subscriptions Collected

Membership dues and subscriptions are assessed and recognized as revenue based on the life of the dues or subscription.

Functional Allocation of Expenses

The costs of providing the various programs and other activities have been summarized on a functional basis in the statement of activities. Accordingly, certain costs have been allocated among the programs and supporting services benefited.

NOTE 2 – ARRANGEMENTS WITH THE UNIVERSITY OF ILLINOIS

The Organization moved its offices to the University of Illinois at the end of 1999. Under an arrangement with the University, the employees of the Organization are employed by the University. The employees' pension and benefits are part of the University's plans. The Organization then reimburses the University quarterly for the cost of its employees.

The University of Illinois holds some cash for the Organization. These "claim on cash" balances were $22,650 and $12,632 as of December 31, 2016 and 2015, respectively. These balances have been included with cash and cash equivalents.

NOTE 3 – UCIRHRP SERVICING ARRANGEMENT TO ADMINISTER BUSINESS

On January 3, 2003, the Organization entered into a servicing arrangement to administer the business of the University Council of Industrial Relations and Human Resources Programs (UCIRHRP). These servicing duties include dues notification and collection, annual meeting arrangement and report preparation, and maintaining a data base and network communications for this separate organization. A separate financial statement is maintained for this company, but UCIRHRP's cash is maintained in the Organization's general bank account. The amount of cash belonging to UCIRHRP in the general account is shown on the Organization's balance sheet as Funds Held for Other Organizations. UCIRHRP's cash balance in the Organization's custody was $6,605 at December 31, 2016 and $5,805 at December 31, 2015. The Organization is allowed to keep 20% of dues collected each year as an administrative fee.

NOTE 4 – CONCENTRATIONS OF CREDIT RISK

Financial instruments that potentially subject the Organization to credit risk consist principally of checking accounts, money markets accounts, and certificates of deposits at financial institutions. However, management continuously monitors the Organization's balances at financial institutions. Deposits held at any financial institutions were fully insured by the Federal Deposit Insurance Corporation (FDIC) at December 31, 2016 or 2015. The "claims on cash" held with the University of Illinois as stated in Note 2 is not insured since it is not held in a financial institution.

Gladys and Walter Gershenfeld Publication Fund

On May 8, 2010, the Organization established a temporarily restricted fund to honor the memory of long-time members and past presidents Walter and Gladys Gershenfeld for the purpose of supporting the Organization's electronic and print publications. The fund will be so named for ten years, at which time the Organization will review the fund's purpose and uses, and whether to continue or modify the fund. The Organization's Executive Board directed that 10% of any gifts received each year are unrestricted and can be used for administrative expenses. As of December 31, 2016 and 2015, these temporarily restricted funds totaled $24,550 and $23,298, respectively.

NOTE 6 – INVESTMENTS

Financial Accounting Standards Board ("FASB") Codification 820, Fair Value Measurements, establishes a framework for measuring fair value. That framework provides a fair value hierarchy that prioritizes the inputs to valuation techniques used to measure fair value. The hierarch gives the highest priority to unadjusted quoted prices in active markets for identical assets or liabilities (Level 1 measurements) and the lowest priority to unobservable inputs (Level 3 measurements). The three levels of the fair value hierarchy under FASB Codification 820 are described below:

Level 1 – Inputs to the valuation methodology are based on unadjusted quoted prices for identical assets or liabilities in active markets that the Organization has the ability to access.

Level 2 – Inputs to the valuation methodology include quoted prices for similar assets or liabilities in active markets, quoted prices for identical or similar assets or liabilities in inactive markets, inputs other than quoted prices that are observable for the asset or liability, and inputs that are derived principally from or corroborated by observable market data by correlation or other means. If the asset or liability has a specified (contractual) term, the Level 2 input must be observable for substantially the full term of the asset or liability.

Level 3 – Inputs to the valuation methodology are unobservable and significant to the fair value measurement.

The asset's and liability's fair value measurement level within the fair value hierarchy is based on the lowest level of any input that is significant to the fair value measurement. Valuation techniques used need to maximize the use of observable inputs and minimize the use of unobservable inputs.

The following is a description of the valuation methodologies used for assets measured at fair value. There have been no changes in the methodologies used at December 31, 2016 and 2015.

Intermediate Term Bond Funds: Valued at the NAV of shares held by the Organization at year-end based on readily determinable fair values, which are published daily and are the basis for current transactions.

The preceding methods described may produce a fair value calculation that may not be indicative of net realizable value or reflective of future fair values. Furthermore, although the Organization believes its valuation methods are appropriate and consistent with other market participants, the use of different methodologies or assumptions to determine the fair value of certain financial instruments could result in different fair value measurements at the reporting date.

The following table sets forth by level, within the fair value hierarchy, the Organization's assets at fair value as of December 31, 2016 and 2015:

Assets at Fair Value as of December 31, 2016

	Level 1	Level 2	Level 3	Total
Intermediate Term Bond Funds	$23,796	-	-	$23,796

Assets at Fair Value as of December 31, 2015

	Level 1	Level 2	Level 3	Total
Intermediate Term Bond Funds	$23,145	-	-	$23,145

Net investment earnings for the years ended December 31, 2016 and 2015 are summarized as follows:

	2016	2015
Dividends	$ 618	$ 622
Capital Gains Distributions	95	79
Net Unrealized Gains (Losses)	(62)	(412)
Total Investment Income	$ 651	$ 289

This investment income, including unrealized gains and losses, are being reported on the statement of activity as investment income.

NOTE 7 – PROPERTY AND EQUIPMENT

As of December 31, 2016, Property and Equipment consists of:

Furniture and Equipment	$ 12,952
Less: Accumulated Depreciation	(12,676)
Property and Equipment, Net	$ 276

As of December 31, 2015, Property and Equipment consists of:

Furniture and Equipment	$ 14,551
Less: Accumulated Depreciation	(13,703)
Property and Equipment, Net	$ 848

Depreciation expense for the years ended December 31, 2016 and 2015 was $572 and $739, respectively.

NOTE 8 – ACCRUED EXPENSES

As of December 31, 2016, Accrued Expenses consist of:

Accrued Payroll	$ 5,318
Accrued Benefits/Taxes	1,653
Accrued Vacation	13,453
Total	$ 20,424

As of December 31, 2015, Accrued Expenses consist of:

Accrued Payroll	$ 4,073
Accrued Benefits/Taxes	1,352
Accrued Vacation	10,132
Total	$ 15,557

NOTE 9 – COMMITMENTS

On February 26, 2013, the Organization signed a 12 month contract from March 1, 2013 to February 28, 2014 with a company to provide services to manage the EPRN website and provide content for the Organization's website. The contract states that the Organization will pay this company $2,400 per month for a total of $28,800. The Agreement automatically renews for consecutive 12 month terms, unless the Agreement is terminated by one or both of the parties as set forth herein. Beginning the second cycle and continuing until termination or renegotiation, the Organization will increase the base rate every year by two percent (2%) over the previous year's base rate. For example: $2,448 per month for twelve months starting March 1, 2014; $2,497 per month for twelve months starting March 1, 2015; $2,546 per month for twelve months starting March 1, 2016; and so on.

The Organization has signed contracts with hotels in Minneapolis, Anaheim, and Baltimore for the 2016, 2017, and 2018 Annual Meetings. If these contracts were cancelled at December 31, 2016 and 2015, the Organization would have owed $128,000 and $109,521, respectively, to the Minneapolis, Anaheim, and Baltimore hotels.

NOTE 10 – PENSION PLAN AND RETIREMENT CONTRIBUTIONS

The University of Illinois (the University) contributes to the State Universities Retirement System of Illinois (SURS), a cost-sharing multiple-employer defined benefit plan with a special funding situation whereby the State of Illinois (the State) makes substantially all actuarially determined required contributions on behalf of the participating employers. SURS was established July 21, 1941 to provide retirement annuities and other benefits for staff members and employees of state universities, certain affiliated organizations, and certain other state educational and scientific agencies and for survivors, dependents, and other beneficiaries of such

NOTE 5 – TEMPORARILY RESTRICTED NET ASSETS

Susan C. Eaton Scholar-Practitioner Memorial Fund

The Organization set up a memorial fund in honor of an author of a 1998 "Perspectives on Work" article and active young the Organization member, who died on December 30, 2003. At its June 1, 2004 meeting, the Organization approved the establishment of an annual Susan C. Eaton Scholar-Practitioner award and grant to be paid to one or more qualified scholar researchers in even-numbered years or practitioners in odd-numbered years doing research in the labor and employment relations or related field. The Organization's Executive Board directed that 10% of any gifts received each year are unrestricted and can be used for administrative expenses. As of December 31, 2016 and 2015, these temporary restricted funds totaled $32,413 and $32,040, respectively.

Kochan-Sleigh Best Dissertation Award Fund

On March 27, 2006, the Organization set up a temporarily restricted fund with contributions from a member and matching contributions from General Electric. The fund is restricted for a minimum of fifteen years and may be used to pay for the $1,000 annual best dissertation award and plaque. After the fifteen year period, the fund may be continued or the amount remaining in the fund may be released and become available to the Organization for unrestricted purposes. The Organization's Executive Board directed that 10% of any gifts received each year are unrestricted and can be used for administrative expenses. As of December 31, 2016 and 2015, these temporary restricted funds totaled $41,941 and $43,090, respectively.

Jim Scoville Award Fund

On January 9, 2009, the Organization set up a temporarily restricted fund with a $10,000 contribution from the University of Minnesota's Industrial Relations Center to honor a member and retiring professor. The fund was established to pay an annual award of $500 and a plaque for best paper on international and comparative employment issues. The Organization's Executive Board directed that 10% of any gifts received each year are unrestricted and can be used for administrative expenses. As of December 31, 2016 and 2015, these temporarily restricted funds totaled $5,251 and $6,455, respectively.

John T. Dunlop Public Policy Fund

On May 8, 2010, the Organization established a temporarily restricted fund to subsidize a named John T. Dunlop Public Policy Session at future the Organization National Policy Forums and/or at the Organization annual meetings. The Organization's Executive Board directed that 10% of any gifts received each year are unrestricted and can be used for administrative expenses. This fund will be so named for five or ten years, at which time the Organization's Executive Board will review the fund to determine whether to continue or modify the fund. As of December 31, 2016 and 2015, these temporary restricted funds totaled $20,657 and $19,809, respectively.

employees. SURS is considered a component unit of the State of Illinois' financial reporting entity and is included in the State's financial reports as a pension trust fund. SURS is governed by Section 5/15, Chapter 40 of the *Illinois Compiled Statutes*. SURS issues a publicly available financial report that includes financial statements and required supplementary information. That report may be obtained by accessing the website at www.SURS.org.

Benefits Provided. A traditional benefit plan was established in 1941. Public Act 90-0448 enacted effective January 1, 1998, established an alternative defined benefit program known as the portable benefit package. The traditional and portable plan Tier 1 refers to members that began participation prior to January 1, 2011. Public Act 96-0889 revised the traditional and portable benefit plans for members who begin participation on or after January 1, 2011, and who do not have other eligible Illinois reciprocal system services. The revised plan is referred to as Tier 2. New employees are allowed 6 months after their date of hire to make an irrevocable election. A summary of the benefit provisions can be found in the SURS' comprehensive annual financial report (CAFR) Notes to the Financial Statements.

Contributions. The State of Illinois is primarily responsible for funding SURS on behalf of the individual employers at an actuarially determined amount. Public Act 88-0593 provides a Statutory Funding Plan consisting of two parts: (i) a ramp-up period from 1996 to 2010 and (ii) a period of contributions equal to a level percentage of the payroll of active members of the System to reach 90 percent of the total Actuarial Accrued Liability by the end of fiscal year 2045. Employer contributions from "trust, federal, and other funds" are provided under Section 15-155(b) of the Illinois Pension Code and require employers to pay contributions which are sufficient to cover the accruing normal costs on behalf of applicable employees. The employer normal cost is currently 12.54 percent of employee payroll. The normal cost is equal to the value of current year's pension benefit and does not include any allocation for the past unfunded liability or interest on the unfunded liability. Plan members are required to contribute 8.0 percent of their annual covered salary. The contribution requirements of plan members and employers are established and may be amended by the Illinois General Assembly. During the years ended December 31, 2016, 2015 and 2014, the Organization reimbursed the University of Illinois $10,388, $13,231, and $14,743, respectively, for the employer's payments to SURS for eligible employees.

In addition to providing pension benefits, the State of Illinois provides certain health, dental and life insurance benefits to annuitants. This includes annuitants of the Organization. Substantially all State employees, including the Origination's employees, may become eligible for postemployment benefits if they eventually become annuitants. Health and dental benefits include basic benefits for annuitants under the State's self-insurance plan and insurance contracts currently in force. Life insurance benefits for annuitants under age 60 are equal to their annual salary at the time of retirement; life insurance benefits for annuitants age 60 or older are limited to $5,000 per annuitant. Currently, the State does not segregate payments made to annuitants from those made to current employees for health, dental and life insurance benefits. These costs are funded by the State and are not an obligation of the Organization.

Employees of the Organization may also elect to participate in several tax deferred annuity plans and defined contribution plans. These are single employer plans under which benefits are provided to participating employees through contracts issued to each individual. Participation and the level of employee contributions are voluntary. The Organization is not required to make contributions.

LABOR AND EMPLOYMENT RELATIONS ASSOCIATION
SCHEDULE OF FUNCTIONAL EXPENSES
FOR THE YEAR ENDED DECEMBER 31, 2016

		MEETINGS			PUBLICATIONS			SUPPORTING SERVICES		
	General	Annual Conference	ASSA Meeting	Other Meetings	Perspectives	Research Volume	Other Publications	Management & General	Membership Development	Totals
Compensation	$ 76,785	$ -	$ -	$ -	$ -	$ -	$ -	$ 17,382	$ -	$ 94,167
Payroll Taxes & Fringes	26,979	-	-	-	-	-	-	7,533	-	34,512
Contract Labor	52,333	-	-	-	-	-	-	-	-	52,333
Depreciation	-	-	-	-	-	-	-	572	-	572
Loss on Disposal of Fixed Assets	-	-	-	-	-	-	-	-	-	-
Insurance	-	-	-	-	-	-	-	3,996	-	3,996
Bank Charges	-	-	-	-	-	-	-	480	-	480
Promotion	-	5,000	-	-	-	-	-	-	5,143	10,143
Fulfillment	-	-	-	-	-	-	-	11,404	-	11,404
Postage and Freight	-	384	42	-	1,516	1,377	-	1,554	-	4,873
Accounting/Auditing	-	-	-	-	-	-	-	4,865	-	4,865
Printing and Production	-	2,204	74	-	5,068	6,854	102	-	-	14,302
Services	-	-	-	-	9,073	7,238	-	-	-	16,311
Other Publication Costs	-	-	-	-	69	-	-	-	-	69
Meals/Receptions	-	51,927	8,329	213	-	-	-	-	-	60,469
Travel	-	2,445	310	967	-	-	-	-	-	3,722
Other Meeting Expenses	-	18,459	1,819	553	-	-	-	-	-	20,831
Computer Supplies/Services	-	-	-	-	-	-	-	88	-	88
Website	-	-	-	-	-	-	-	1,281	-	1,281
Office Supplies	-	-	-	-	-	-	-	883	-	883
Student & Member Awards	-	-	-	-	-	-	-	-	4,794	4,794
Fundraising Expense	-	-	-	-	-	-	-	-	2,117	2,117
Duplicating Expense	-	-	-	-	-	-	-	1,150	-	1,150
Telephone & Fax	-	-	-	-	-	-	-	426	-	426
Chapter Expenses	-	-	-	-	-	-	-	-	132	132
Dues	-	-	-	-	-	-	-	-	-	-
Other Committee Expenses	-	-	-	-	-	-	-	-	2,572	2,572
Miscellaneous Office	-	-	-	-	-	-	-	45	-	45
	$ 156,097	$ 80,419	$ 10,574	$ 1,733	$ 15,726	$ 15,469	$ 102	$ 51,659	$ 14,758	$ 346,537

LABOR AND EMPLOYMENT RELATIONS ASSOCIATION
SCHEDULE OF FUNCTIONAL EXPENSES
FOR THE YEAR ENDED DECEMBER 31, 2015

		MEETINGS			PUBLICATIONS			SUPPORTING SERVICES		
	General	Annual Conference	ASSA Meeting	Other Meetings	Perspectives	Research Volume	Other Publications	Management & General	Membership Development	Totals
Compensation	$ 96,665	$ -	$ -	$ -	$ -	$ -	$ -	$ 17,922	$ -	$ 114,587
Payroll Taxes & Fringes	38,551	-	-	-	-	-	-	7,113	-	45,664
Contract Labor	31,021	-	-	-	-	-	-	-	-	31,021
Depreciation	-	-	-	-	-	-	-	739	-	739
Loss on Disposal of Fixed Assets	-	-	-	-	-	-	-	-	-	-
Insurance	-	-	-	-	-	-	-	3,843	-	3,843
Bank Charges	-	-	-	-	-	-	-	1,851	-	1,851
Promotion	-	2,906	149	-	-	-	-	-	3,397	6,452
Fulfillment	-	-	-	-	-	-	-	7,664	-	7,664
Postage and Freight	-	300	-	-	2,214	1,263	-	1,418	-	5,195
Accounting/Auditing	-	-	-	-	-	-	-	4,850	-	4,850
Printing and Production	-	2,490	139	-	7,505	7,338	43	-	-	17,515
Services	-	-	-	-	12,221	9,307	-	-	-	21,528
Other Publication Costs	-	-	-	-	1,390	-	323	-	-	1,713
Meals/Receptions	-	66,513	3,464	-	-	-	-	-	-	69,977
Travel	-	4,692	976	-	-	-	-	-	-	5,668
Other Meeting Expenses	-	7,269	410	11	-	-	-	-	-	7,690
Computer Supplies/Services	-	490	-	-	-	-	-	2,861	-	3,351
Website	-	-	-	-	-	-	-	1,900	-	1,900
Office Supplies	-	-	-	-	-	-	-	77	-	77
Student & Member Awards	-	-	-	-	-	-	-	-	3,042	3,042
Fundraising Expense	-	-	-	-	-	-	-	-	-	-
Duplicating Expense	-	-	-	-	-	-	-	315	-	315
Telephone & Fax	-	-	-	-	-	-	-	383	-	383
Chapter Expenses	-	-	-	-	-	-	-	-	-	-
Dues	-	-	-	-	-	-	-	-	-	-
Other Committee Expenses	-	-	-	-	-	-	-	-	273	273
Miscellaneous Office	-	-	-	-	-	-	-	43	-	43
	$ 166,237	$ 84,660	$ 5,138	$ 11	$ 23,330	$ 17,908	$ 366	$ 50,979	$ 6,712	$ 355,341

LERA gratefully acknowledges the continuing support of its Sustaining Sponsors and Organizational Members

SUSTAINING SPONSORS 2017

Ford Motor Company
Kaiser Permanente

ANNUAL MEMBERS 2016–2017

American Federation of Teachers
Berkeley Research Group, Labor and Employment Practice
Boston University, Questrom School of Business
Central Alabama Labor Federation
Consortium for Educational Change
Cornell University, Scheinman Institute on Conflict Resolution
Cornell University, School of Industrial and Labor Relations
Federal Mediation and Conciliation Service
Florida International University, College of Business
International Association of Machinists and Aerospace Workers
Jobs with Justice Education Fund
Massachusetts Institute of Technology, Sloan School of Management
Pennsylvania State University, School of Labor and Employment Relations
Rollins College, Department of Business & Social Entrepreneurship
Rutgers School of Management and Labor Relations
Saint Joseph's University, Erivan K. Haub School of Business
Society for Human Resources Management
Tennessee State Employees Association
The Dannon Company
UAW National CAP
United Steelworkers of America
University of California, Berkeley, Center for Labor Research and Education
University of California, Berkeley, Institute for Research on Labor and Employment
University of California, Los Angeles, Institute for Research of Labor and Employment
University of California, Los Angeles, Anderson School of Management
University of Illinois at Urbana-Champaign, School of Labor and Employment Relations
University of Minnesota, Center for Human Resources and Labor Studies
University of Newcastle
West Virginia University, College of Business and Economics

For more information about Organizational Membership with the LERA, please visit the LERA website: https://lera.memberclicks.net/organizational-memberships

ALPHABETICAL LIST OF AUTHORS